# SpringerBriefs in Geography

SpringerBriefs in Geography presents concise summaries of cutting-edge research and practical applications across the fields of physical, environmental and human geography. It publishes compact refereed monographs under the editorial supervision of an international advisory board with the aim to publish 8 to 12 weeks after acceptance. Volumes are compact, 50 to 125 pages, with a clear focus. The series covers a range of content from professional to academic such as: timely reports of state-of-the art analytical techniques, bridges between new research results, snapshots of hot and/or emerging topics, elaborated thesis, literature reviews, and in-depth case studies.

The scope of the series spans the entire field of geography, with a view to significantly advance research. The character of the series is international and multidisciplinary and will include research areas such as: GIS/cartography, remote sensing, geographical education, geospatial analysis, techniques and modeling, landscape/regional and urban planning, economic geography, housing and the built environment, and quantitative geography. Volumes in this series may analyze past, present and/or future trends, as well as their determinants and consequences. Both solicited and unsolicited manuscripts are considered for publication in this series.

SpringerBriefs in Geography will be of interest to a wide range of individuals with interests in physical, environmental and human geography as well as for researchers from allied disciplines.

More information about this series at http://www.springer.com/series/10050

Jarosław Działek · Wojciech Biernacki
Roman Konieczny · Łukasz Fiedeń
Paweł Franczak · Karolina Grzeszna
Karolina Listwan-Franczak

# Understanding Flood Preparedness

Flood Memories, Social Vulnerability
and Risk Communication in Southern Poland

 Springer

Jarosław Działek
Institute of Geography and Spatial
Management, Faculty of Geography
and Geology
Jagiellonian University
Kraków, Poland

Wojciech Biernacki
Faculty of Tourism and Leisure
University of Physical Education in Krakow
Kraków, Poland

Roman Konieczny
Institute of Meteorology and Water
Management
National Research Institute
Kraków, Poland

Łukasz Fiedeń
Institute of Geography and Spatial
Management, Faculty of Geography
and Geology
Jagiellonian University
Kraków, Poland

Paweł Franczak
Institute of Geography and Spatial
Management, Faculty of Geography
and Geology
Jagiellonian University
Kraków, Poland

Karolina Grzeszna
Independent Researcher
Kraków, Poland

Karolina Listwan-Franczak
Institute of Geography and Spatial
Management, Faculty of Geography
and Geology
Jagiellonian University
Kraków, Poland

ISSN 2211-4165                    ISSN 2211-4173    (electronic)
SpringerBriefs in Geography
ISBN 978-3-030-04593-7           ISBN 978-3-030-04594-4    (eBook)
https://doi.org/10.1007/978-3-030-04594-4

Library of Congress Control Number: 2018962381

This Springer imprint is published by the registered company Springer Nature Switzerland AG
The registered company address is: Gewerbestrasse 11, 6330 Cham, Switzerland

# Contents

# Chapter 1
# Introduction

Floods are among the natural hazards, which cause the greatest financial losses in Europe (EEA 2010). In addition, they have serious adverse social consequences and implications for human health. Over the last 30 years, catastrophic floods affected the Central and Eastern Europe several times (Raška 2015; Gvoždíková and Müller 2017). Meanwhile, the paradigm, which prevails in European post-socialist countries, remains based on flood protection, where the focus is on the physical course of the phenomena that put humans at risk. It relies on using scientific and engineering expertise to control the forces of nature by employing hydrotechnical solutions ('moving water away from people'). This is a typical top-down approach where the authorities and experts take decisions and residents of flood-prone areas are required to comply with them (Fordham et al. 2013). In addition, as is observed by Raška (2015), both decision-makers and the public perceive floods as a temporary issue, which tends to be addressed only during extreme events and in their wake. This is reflected by the associated risk communication processes, which tend to focus on disaster communication, with limited flood risk management at later stages.

The new philosophy of flood risk management, which takes greater account of the bottom-up perspective, namely, the voice of vulnerable communities in bilateral communication processes, is emerging very slowly in this part of Europe. For example, the approach to flood risk mitigation that limits exposure to floods, or in other words 'moves people away from water', through appropriate spatial planning, comes across a number of barriers and also encounters resistance from the populations exposed to floods. On the one hand, the obstacles result from the planning tradition in the post-socialist countries, and on the other, from the perception of private ownership as a fundamental right after the decades of centralised control of property in the pre-1989 period. The concept of vulnerability, in particular, social vulnerability, which provides decision-makers with information on the extent to which communities at risk of natural disasters are capable of coping with them, has not been widely assimilated by flood management stakeholders (Hillhorst and Bankoff 2004; Mercer et al. 2008; Höppner et al. 2010; *Understanding the risks* … 2011; Żelaziński 2011; Bryndal 2014; Raška 2015; Konieczny et al. 2017). Similarly, the concept of shared responsi-

© The Author(s), under exclusive licence to Springer Nature Switzerland AG 2019    1
J. Działek et al., *Understanding Flood Preparedness*, SpringerBriefs in Geography,
https://doi.org/10.1007/978-3-030-04594-4_1

bility for flood mitigation has not yet emerged widely in public discourse; however, one can imagine the reluctant attitude of Polish citizens towards such a shift in flood risk management policies as is happening in those countries that have started such a debate (Mees et al. 2016; Henstra et al. 2018).

In this book, our main point of interest is the flood preparedness of households living in flood risk areas. The question of insufficient individual preparation for natural disasters is still an important issue to be explained. In this context, attention is drawn to the impact of the two mechanisms influencing disaster preparedness: risk perception and social vulnerability (Fox-Rogers et al. 2016).

Risk perception studies form a well-developed field of disaster research (Lindell 2013); hence, we wanted to explore the less frequently investigated issue of how a flood event is remembered and perceived in terms of its spatial dimension. Individual and collective images of zones at risk of flooding constructed by members of 'at risk' communities could be one of the factors prompting them to undertake mitigation activities. Cognitive mapping with the use of sketch maps drafted twice by the same group of respondents has demonstrated an important variation of individual perceptions, however, with a common tendency to underestimate size of flood zones additionally decreasing with time.

Fading social memories of disasters in the spatial dimension could be counterweighed with effective risk communication in land-use planning documents (de Moel et al. 2009; Neuvel and van der Knaap 2010; Głosińska 2014). This has encouraged us to examine how flood experiences have been translated into official planning documents in more detail. Analysis of the legal background and practice of preparing land-use plans in those municipalities that were affected by floods has shown that there is no clear and unified procedure of delimiting floodplains and introducing them in official records. Additionally, our study concludes that in some cases, especially in mountainous areas, the width of flooded area was larger than delimited in documents. Miscommunication of hazard in land-use planning could affect flood risk perception and result in higher vulnerability of local communities.

Land-use plans and flood risk maps are becoming more and more easily accessible on the Internet, forming part of developing online risk communication tools. They are a source both of hope of higher availability of information about various risks, and of fear of spreading misinformation (Roth and Brönnimann 2013). We have decided to focus on two groups of flood-related content generators: national and regional agencies and local municipalities on one side, and social media users, notably the YouTube video sharing portal. On one hand, we have discovered that official sources offer dispersed information about hazards and their mitigation. On the other hand, user-generated content on YouTube includes a number of flood recordings creating 'social archives' of flood memories, but without giving a broader context of the events they picture. At the same time, official institutions do not use this channel for risk communication and education purposes, losing the opportunity to use the growing interest of Internet users searching for information on social media.

Another major issue related with flood preparedness is the notion of social vulnerability that denotes the ability of individuals and groups to foresee a disaster, react to it and endure it, and finally recover after it happens (Blaikie et al. 1994). It refers to different stages of risk management. Our interest was concentrated on the pre-disaster phase, when households decide whether to undertake any preparatory activities. We test the concept of social vulnerability by assuming that various social and economic factors hinder, for different reasons, individual or collective capacities to prepare themselves. Our main observation is that socio-economic characteristics explain better households' flood preparedness than their demographic situation. Still, one group, that is, the elderly, proves to be more vulnerable even if we consider their general more difficult socio-economic situation.

The above considerations provoked the idea for the research presented in this book. At the heart of it are communities inhabiting flood-prone areas, with a focus on the period between floods (pre-disaster phase of flood risk management). The purpose is to identify the determinants which stimulate people to prepare for a flood several years after the disaster, rather than immediately after it, when they are more motivated.

The assumption underlying this book is that introducing a new approach to risk management—one in which vulnerable communities participate in the process and are co-responsible for it—requires determining their situation on a local scale, including numerous limitations. This would be helpful in developing policies oriented to 'learning how to live with the flood'. Thus, our study focuses on four interrelated research questions, which are presented in the successive parts of this book:

1. How does the social memory of a flood evolve several years after the disaster? In particular, how is the spatial image of a flood formed and how does it change over time? (Chap. 3)
2. What are the differences in flood preparedness among residents of flood-prone areas? What is the role in this context of the socio-economic drivers of social vulnerability to flooding? (Chap. 4)
3. What is the role of the Internet in communicating flood risk in vulnerable areas? What is the attitude of the different stakeholders to this process? (Chap. 5)
4. How is flood risk communicated in land-use planning documents? (Chap. 6)

A later part of this chapter introduces the research design adopted in our study, in particular, the set of research methods that enabled us to give answers to the above questions. It briefly presents all the study areas chosen in this research project and the floods that have affected them. The broader meteorological and hydrological context of these disasters is discussed in Chap. 2. Literature relevant to each research question is reviewed in their respective chapters to give a clear picture of the current state of research in each of the fields that, although related to each other, constitute independent issues. The final part of the book brings together the conclusions from all the chapters and reveals connections between them.

The book was written as part of research carried out within the research project on "Socio-economic factors of social vulnerability to floods with a special focus on the role of communication" financed with a grant awarded by the National Science Center, Poland, Grant agreement no. UMO-2012/05/D/HS4/01328.

## 1.1  Research Design

The study comprises five regions of southern Poland: Dolnośląskie, Opolskie, Śląskie, Małopolskie and Podkarpackie (Fig. 1.1), which were selected on account of their being at the greatest risk of flooding at a countrywide scale. They were affected by the three major regional floods which hit Poland in 1997, 2001 and 2010 (Kowalewski 2006; Kundzewicz et al. 2010; Halama 2013a, b; *Powódź … 2013*; *Zagrożenia okresowo … 2013*). Smaller, often local floods are happening more often in various parts of these Polish regions. Both flooding events that occurred during these large floods and smaller localised river overflows were studied in this research. The high flood risk in the Southern part of the country is due to the increased flood potential of the rivers of the upper Vistula basin and of the tributaries of the Odra River in the Sudetes (Magnuszewski and Porczek 2015), their high population density, particularly in the Carpathian valleys, and the growing pressure to develop the floodplains within those areas.

Several complementary quantitative and qualitative research methods were used to answer the above questions. Publicly available information, including meteorological and hydrological data acquired from IMGW-PIB (Polish Institute of Meteorology and Water Management—National Research Institute) and other documentation on recent regional and local floods (from the end of the 1990s to the beginning of the 2010s) and their consequences obtained specifically for the purpose of the study (e.g. from regional authorities), was used to select case studies for field research. The criteria for their selection included the size of the locality (from rural areas to medium-sized towns), flood frequency and the nature of the flood.

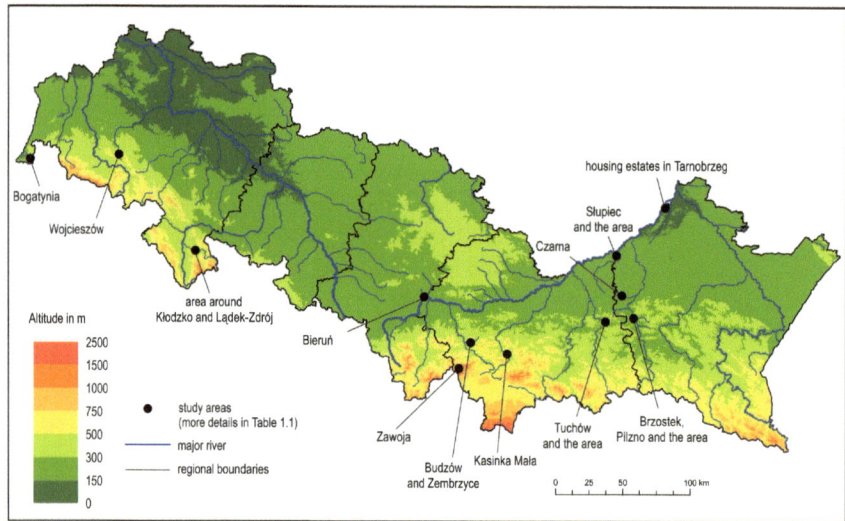

**Fig. 1.1**  Study areas. *Source* Own elaboration

Four main research methods were used to address the four main research questions with different sets of study areas according to the specificities of each approach. Cognitive mapping of flooded areas and interviews with the local population were used to uncover patterns of the spatial dimension of flood memories. Since it was the most time- and work-consuming method, only three study areas were selected with flood zones of a relatively small size (Table 1.1). A questionnaire survey was used in eight areas of a larger size to get information about the flood preparedness of at-risk populations and uncover drivers of social vulnerability influencing their willingness to undertake flood mitigation activities. YouTube videos of ten flood events uploaded by social media users from all the areas studied during the cognitive mapping and survey were analysed to understand the role of the Internet in flood risk communication. A wider range of Internet materials (national and regional authorities and emergency services, websites of municipalities with a high flood risk) were analysed for flood-related content to bring a broader understanding of online flood risk communication. Finally, land-use planning documents were examined in six study areas—due to the work demanding character of this study, eight localities and respective river catchments of different sizes were selected to recognise the role of planning documents in flood risk communication. All research methods are described in more detail in a later part of this introductory chapter.

Altogether, 12 study areas from various areas of southern Poland (Fig. 1.1, Table 1.1) helped us to better understand various facets of flood preparedness and their relations with flood memories, social vulnerability and flood risk communication.

## 1.2 Summary of Flood Events

After a relatively dry period in the 1980s and beginning of the 1990s, three major regional floods and a whole series of local floods were recorded in southern Poland over the last 25 years. In July 1997, there was a catastrophic flood which particularly affected the Odra basin, and which, due to the size of the resulting losses, was called the 'Millennium Flood'. It was caused by intense rainfall with the 3-day sum of precipitation exceeding 450 mm, and the maximum daily sums reaching 180 mm. One of the largest losses recorded at that time occurred in the Kłodzko Valley. The maximum flow of the Nysa Kłodzka in Kłodzko amounted to 1340 m$^3$ s$^{-1}$ (with an average flow of 13.1 m$^3$ s$^{-1}$). The economic losses over the whole country were estimated at PLN 12.8 billion, and the flood took the life of 55 people (the highest death toll and losses since the 1934 and 1970 floods).

The second large flood in southern Poland occurred 4 years later, also in July. It was caused by intense, long-lasting (for almost a month) rainfall, with overlapping repeating several-hour long rainstorms causing local flash floods. The highest precipitation total was recorded in the Skawa basin, with a daily sum of precipitation in Maków Podhalański of 190.8 mm. At that time, great damage was caused by streams flowing from the Makowska Mountain (in Budzów and Zembrzyce, and in Maków

**Table 1.1** Basic characteristics of study areas in chronological order of flood and the research methods used in each of them

| Study areas | Area characteristic | Year of flooding studied during field research | Flood memories mapping | Questionnaire survey | YouTube videos study | Land-use planning documents study |
|---|---|---|---|---|---|---|
| Area around Kłodzko and Lądek-Zdrój | Five villages (population of 0.5–3 thousand) in a mountain valley, part of Kłodzko Basin (Sudety Mountains). Territory acquired after 1945 | July 1997 (R)* | | x | x | x |
| Budzów and Zembrzyce | Three villages (population of 1–3 thousand) in a Maków Beskids mountain valley (Carpathian Mountains) | July 2001 (R) | | x | x | x |
| Czarna | Large village (2.7 thousand) in the Tarnów Plateau, surrounded by farmland and forest | June 2009 (L) | x | | x | |
| Brzostek, Pilzno and the area | Two small towns (3 and 5 thousand) and some neighbouring villages (0.5–1 thousand) at the boundary between Ciężkowice and Strzyżów Foothills (Carpathian Mountains) | May and June 2010 (R) | | x | x | |
| Tuchów and the area | Small town (7 thousand) and two neighbouring villages (each ca. 1 thousand) on the Ciężkowice and Rożnów Foothills (Carpathian Mountains) | May and June 2010 (R) | | x | x | x |

(continued)

**Table 1.1** (continued)

| Study areas | Area characteristic | Year of flooding studied during field research | Flood memories mapping | Questionnaire survey | YouTube videos study | Land-use planning documents study |
|---|---|---|---|---|---|---|
| Bieruń | Medium-sized town (20 thousand) in the Upper Silesian Oświęcim Basin; nearly a coal mine company town | May and June 2010 (R) | x | x | x | |
| Słupiec and the area | Five villages (0.5–1 thousand each) in a broad river valley plain within the Vistula River Bank Lowland (part of Sandomierz Basin) straddling two voivodeships | May and June 2010 (R) | | x | x | |
| Housing estates in Tarnobrzeg | Four suburban housing estates of this medium-sized town (48 thousand), former villages incorporated in the town in 1975, but retaining their separate nature; ca. 0.5–1.5 thousand; Vistula River Bank Lowland (part of Sandomierz Basin) | May and June 2010 (R) | | x | x | |
| Bogatynia | A mid-size town (18 thousand) close to Poland's border with the Czech Republic and Germany, a territory acquired in 1945, a brown-coal mine town | August 2010 (L) | | x | x | x |

(continued)

**Table 1.1**  (continued)

| Study areas | Area characteristic | Year of flooding studied during field research | Flood memories mapping | Questionnaire survey | YouTube videos study | Land-use planning documents study |
|---|---|---|---|---|---|---|
| Wojcieszów | Small town (3.8 thousand) within a forest area in the borderland between the Kaczawa Foothills and the Kaczawa Mountains, located in the historical region of the Lower Silesia, which was incorporated into Poland in 1945 | July 2012 (L) | x | | x | |
| Zawoja | Large village (6.5 thousand) in mountainous area of the Żywiec and Maków Beskids | May 2014 (L) | | | | x |
| Kasinka Mała | Large village (3.4 thousand) in mountainous area of Island Beskids | August 2014 (L) | | | | x |

*Notes* * (L)—isolated local flood, (R)—one of the local events during large regional floods

Podhalański). In the profile located at the dam on the Skawa in Świnna Poręba that existed at that time, the maximum flow amounted to as much as $1019 \text{ m}^3 \text{ s}^{-1}$. Losses in the Vistula basin amounted to approximately PLN 4.2 billion.

The last of the major floods under study occurred in 2010, when three flood waves formed from May to September in the south of Poland. In the western part of the upper Vistula basin, the sum of rainfall in May amounted to almost 590 mm. This resulted in the formation of a flood wave on the Vistula and its tributaries. The maximum flow of the Vistula in Szczucin amounted to $5270 \text{ m}^3 \text{ s}^{-1}$. In turn, in the basin of the Biała Tarnowska and Wisłoka, the highest flood wave was formed in June. During the flood, the largest backwater was created within the Sandomierz Basin in Tarnobrzeg. In turn, the maximum flow of the Vistula in Nowy Bieruń amounted to $846 \text{ m}^3 \text{ s}^{-1}$, reaching the probability of occurrence of the so-called millennial water. Ground subsidence caused by coal mining activities in this area increased the size of the flooded zone in Bieruń.

In the other study catchments, the floods were mainly caused by local but very intense rainstorms, usually lasting up to several hours. The daily rainfall totals reached up to 100 mm; however, during the most intense episodes, hourly precipitation was often higher than 30 mm per hour. This resulted in flash floods, with high flood waves forming just a few hours after the heavy rainfall. Such events were examined in Bogatynia (2010), Wojcieszów (2012), Kasinka Mała and Zawoja (both in 2014).

## 1.3  Research Methods

### 1.3.1  Cognitive Mapping

The studies of the evolution of the spatial image of a flood (research question 1) took the form of panel research conducted in three localities affected by floods over a period of several years preceding the field surveys: the village of Czarna (Podkarpackie Region)—flood in May 2009, the town of Bieruń (Śląskie Region)—flood in May 2010, and the town of Wojcieszów (Dolnośląskie Region)—flood in July 2012. They are small localities, with a population of at most a dozen or so thousand, where flood occurred in the last 10 or more years, and never reoccurred—the authors made sure that the residents' memories always concerned the same event.

The first meetings with the respondents took place in September 2014. The same persons were interviewed again in September 2016. During the first round, 15 or more accounts of the flood were collected as planned, while during the next round, some of the respondents refused to participate. In spite of this, thanks to the surplus interviews from the first round, altogether opinions of 15 respondents were successfully obtained twice. Only in Wojcieszów, the number of refusals was so high that out of the 15 interviews scheduled, only 11 were completed in the second round.

The maps used in the study were based on a 1:6000 orthophotomap, which allowed individual buildings to be recognised. Colour printouts were prepared and then presented to those surveyed during the face-to-face meetings. Using them, each respondent was able to locate easily their own house, the key landmarks and the areas they most frequently visited in the town, and finally to mark, in the presence of the interviewer, the area they believed had been flooded during the last flood. This activity was a pretext for introducing the topic and provided, in most of the cases, a backdrop for the oral account of the flood, which was recorded and analysed. Every effort was taken to ensure that the interviews were as natural as possible, and did not cause unnecessary discomfort to the respondents. Most of them took the form of casual conversations, and—depending on the individual character of the respondent—they lasted from a dozen or so minutes to an hour and a half.

The spatial information collected during the face-to-face interviews with the flood victims was processed using GIS tools. In the first step, analogue data about the respondents' place of residence, their activity and the extent of the flood was digitised to a vector format. The linear and polygonal data was transferred with an accuracy of 1 m onto the areas where landmarks were discernible, and with an accuracy of 3 m, onto the other areas. The areas identified by the respondents as 'flooded' were assigned the value 1, while the remaining areas 0 (Brennan-Horley 2010). The vector data illustrating the flood extent was converted to a raster format (pixel size: $1 \times 1$ m). The last stage of the data conversion process consisted in summing up the values of the pixels representing the same coordinates, but derived from digitised information provided by different persons. The result maps produced showed the number of indications of specific areas of the towns affected by floods according to respondents.

In addition to creating the maps used in the spatial analysis, several indicators were calculated for the needs of quantitative analyses. The metrics established for each respondent included the size of the affected area indicated by the respondent $(km^2)$, the respondent's activity, i.e. the length of roads the person surveyed travels frequently within the locality (km), the distance between the centroid of the area indicated as flooded and the respondent's home.

The spatial extent of the flood established through interviewing was compared to the actual extent of the flood, i.e. the respective inundated areas delimited for the floods in Czarna (2009), Bieruń (2010) and Wojcieszów (2012). The flood areas were determined through field surveys using a GPS receiver, where the reach of the floodwater was identified on the basis of extant signs of the flood, as well as photographs and audiovisual materials. LiDAR data retrieved from the State Geodetic and Cartographic Resources (CODGiK) was also used. Information about the reach of the floodwaters in Wojcieszów was supported by Google Street View material, which was retrieved in July 2012 (several days after the flood). The photographs taken in Wojcieszów show visible traces of the flood along the channel of the Kaczawa River, which crosses the town.

## 1.3.2   Questionnaire Survey

A standardised questionnaire survey of the links between social vulnerability and flood preparedness (research question 2) was conducted in June and July 2014 in eight areas of southern Poland which had suffered from floods over the past dozen or so years preceding the survey, mainly the floods in the summer of 2010. The questions concerned the course of the floods, their consequences and ways of preparation for the floods. They also addressed the socio-economic determinants likely to increase or decrease social vulnerability to a flood. The questionnaire was inspired by research conducted in the catchment area of the Mulde River under the FLOODsite project (Steinführer and Kuhlicke 2007). A wide range of existing indicators from various studies (e.g. Cutter et al. 2003; Birkmann 2006; Steinführer et al. 2009; Tapsell et al. 2010; Kuhlicke et al. 2011; Fekete 2012; Tate 2012; Yoon 2012; Thomas et al. 2013; Rufat et al. 2015) were analysed to provide the most adequate list of possible drivers of social vulnerability. The areas surveyed differed in terms of the size of the localities (villages, small- and medium-sized towns), the frequency of floods (one or more major floods and cases of local flooding) and the nature and course of the flood (flash floods in the mountains and lowland floods). They are located in regions which are very varied in natural, historical, social and economic terms.

A total of 726 households directly affected by the floods were surveyed by trained interviewers recording the answers given directly by the household members during face-to-face interviews. The interviewers tried to reach all the households affected by the floods by conducting surveys at different times of day for several days. The success rate was about 57%—similar to other comparable studies (Steinführer and Kuhlicke 2007). Main reasons for not participating in the survey were both indirect (absence of household members, unoccupied houses) and direct (refusals to partici- pate, household moved to the area after the most recent flood). Within areas with a higher number of households affected, the surveys were conducted in different parts of the flooded areas until at least 100 completed questionnaires were obtained in each of them. Samples from all the areas under study are non-random and cannot be regarded as representative, even though interviewers undertook several attempts to get in contact with household members. However, the use of non-random samples in natural disaster studies is documented (Burke et al. 2010). In a similar manner to other survey studies (Steinführer and Kuhlicke 2007), a higher refusal rate was observed from elderly inhabitants. Additionally, it was not possible to determine the population structure of flooded areas to compare it with the structure of households under study as there are no statistical data available for these very specific areas. Despite the shortcomings, the data gathered during fieldwork can be treated as a valuable source of small-scale data that can be used to understand the mechanisms behind the social vulnerability and flood preparedness of different social groups (Spence and Lachlan 2010). The survey was used to evaluate the self-assessed flood preparedness of households and the presence or absence of factors that are discussed in the literature as factors of social vulnerability. Even though there is a noticeable overuse of quantitative methods to study disasters (Fox-Rogers et al. 2016), they can provide larger sets of data that can be complemented with qualitative methods.

Subjective self-assessment of flood preparedness should be approached with some carefulness as it might be influenced by different factors. However, we consider our respondents to be experts in assessing the situation of their households and the envi- ronment around them, and later on we compare these more subjective opinions with factual statements about their socio-economic situation. We are aware of the limita- tions of conducting surveys after traumatic events and the psychological mechanisms related to them; however, our research was done several years (at least 4 years) after the disaster, so the affected population has a longer time perspective of the events they are referring to, even though these memories could be reinterpreted afterwards (Spence and Lachlan 2010). Being aware of these limitations, as well as a careful consideration of all the information gathered during the survey, allowed us to obtain a better understanding of social vulnerability to flood.

### *1.3.3  Online Risk Communication*

The online content survey, analysed from the perspective of flood risk communication (research question 3), was covered approximately 2 years from the previous major flood in southern Poland. Thus, it can be concluded that the content of websites of the institutions under study, both central and local, refers to a preparatory period 'between floods'.

First of all, the survey looked at websites of central authorities (including the Ministry of the Interior and Administration, Ministry of the Environment, the National Water Management Authority, the Government Centre for Security, the Polish Institute of Meteorology and Water Management, the National Headquarters of the State Fire Service and the Chief Sanitary Inspector), some of which are responsible for key flood risk management and response tasks, while others play a supporting role. The analysis also covered websites of two regional institutions responsible for water and flood risk management, i.e. Regional Water Management Authorities in Kraków and Wrocław.

Two levels of the analysis of the institutions' websites were adopted. The first one is the activity of an institution in relation to selected target groups. Since Polish legislation lacks specific guidelines as to who the different institutions should communicate with and what matters they should address, the study assumes a broad range of potential target groups (from institutions to the general public). An assumption was also made as to the expected level of activity of the institutions vis-à-vis these target groups, which was used as a basis for discussing such activity and the related problems. The second level of analysis included the content published on selected websites of the institutions, which was subdivided according to flood risk management elements, i.e. hazard, exposure, as well as vulnerability of infrastructure and communities to floods.

Second, the availability on the Internet of studies funded by the central budget and the use of online content in textbooks for young people in Poland were analysed. To this end, quantitative and qualitative analyses were carried out of 'nature', 'geography' and 'education for security' textbooks for lower and upper secondary schools. The textbooks included in the analysis were those available on the publishing market in 2016.[1]

Third, at the end of 2014 and in the early 2015, an analysis was conducted of flood-related information available on the official websites of municipalities from the three regions (voivodships) exposed to the highest flood risk, i.e. Dolnośląskie, Małopolskie and Podkarpackie. The study included 344 municipalities which, according to official data, had suffered from at least two of the three major floods that hit those parts of the country in 1997, 2001 and 2010. The review of webpages was not focusing on these three large flood events—this was simply the way to narrow down the number of municipal websites assessed to those that have suffered large flood losses

---

[1]It must be noted that currently Poland is (2017) in the process of a school reform, which involves significant changes to the core curricula used as a basis for the preparation of textbooks for all school levels.

in the two decades prior to the study and where local authorities were assumed to be more mobilised to provide flood risk information. Each website was assessed according to the same procedure. First, an examination was made whether it covers topics related to flood risk before a flood happens (information on potential hazards and how to prepare for them, evacuation routes, previous historic floods and flood risk management documents), during the flood (warnings, information about how it is developing, contact details to crisis management units) and after it has receded (what needs to be done to recover from the flood). Second, the visibility and availability of materials on websites was assessed—whether they are available on the frontpage (highest visibility), whether they can be found after browsing the website, or whether only an embedded search engine needs to be used to find them (lowest visibility).

Finally, the YouTube service was analysed for videos concerning floods in the study areas. Based on the existing methodology (McMillan 2000; Bromberg et al. 2012; Jina and Junghyun 2012; Gradlyan and Baghdasaryan 2013; Luo et al. 2014; Nasim et al. 2014), the categorisation key and sampling principles were established. The names of towns from the 10 study areas were entered in the YouTube search tool, with the word powódź ('flood' in Polish) added. The first 20 search results displayed on the first page (first, with the use of the 'relevance' filter, and then the 'view count' filter) were included in the research sample, excluding, however, videos that were inconsistent with the purpose of the study and the same videos posted repeatedly. A total of 145 films were analysed.

Thanks to its dynamism and freedom of creating new content, the Internet has become one of the key tools of public communication, and as such, it deserves special attention in conducting research on human behaviour and attitudes towards contemporary phenomena (Jemielniak 2013).

### 1.3.4  Analysis of Land-Use Planning Documents

Spatial planning documents from eight localities (Bogatynia, Ołdrzychowice Kłodzkie and Krosnowice near Kłodzko, Budzów and Maków Podhalański, Zawoja, Kasinka Mała and Tuchów) located in six research areas were examined to uncover how flood risk is communicated in official records (research question 4). They were chosen to represent catchments of various sizes (from 2.1 to 920 km$^2$) and orders (from the second to the fifth order) in the upper Vistula and upper and middle Odra basins.

The main aim was to assess the reliability of delimitations of flood-prone areas in municipal records. Flood hazard zones designated in planning documents (in Poland they are represented by a more detailed local spatial development plan—*miejscowy plan zagospodarowania przestrzennego*, and a more general study of conditions and directions of spatial development—*studium uwarunkowań i kierunków zagospodarowania przestrzennego*) were compared with the actual size of floods that had recently occurred in the research areas calculated on the basis of high-water marks. GIS software was used to calculate the width of actual flood zones and flood hazard zones in planning documents for each cross section of the designated river valley every 250 m along its course. Differences between the two zones were analysed and interpreted in each locality.

# References

Birkmann J (ed) (2006) Measuring vulnerability to natural hazards: towards disaster resilient societies. United Nations University Press, Tokyo, New York, Paris

Blaikie P, Cannon T, Davis I, Wisner B (1994) At risk: natural hazards, people's vulnerability, and disaster. Routledge, London

Brennan-Horley C (2010) Creative city mapping: experimental applications of GIS for cultural planning and auditing. University of Wollongong, Wollongong

Bromberg JE, Augustson EM, Backinger CL (2012) Portrayal of smokeless tobacco in YouTube videos. Nicotine Tob Res 14(4):455–462

Bryndal T (2014) Powodzie błyskawiczne w małych zlewniach karpackich—wybrane aspekty zarządzania ryzykiem powodziowym. Ann Univ Paedagogicae Cracoviensis Stud Geogr 7(170):69–80

Burke JA, Spence PR, Lachlan KA (2010) Crisis preparation, media use, and information seeking during Hurricane Ike: lessons learned for emergency communication. J Emergency Manage 8(5):27–37

Cutter SL, Boruff BJ, Shirley WL (2003) Social vulnerability to environmental hazards. Soc Sci Q 84(1):242–261

de Moel H, van Alphen J, Aerts JCJH (2009) Flood maps in Europe—methods, availability and use. Nat Hazards Earth Syst Sci 9:289–301

EEA (2010) Mapping the impacts of natural hazards and technological accidents in Europe. An overview of the last decade. EEA technical report no 13/2010. European Environment Agency, Copenhagen

Fekete A (2012) Spatial disaster vulnerability and risk assessments: challenges in their quality and acceptance. Nat Hazards 61:1161–1178

Fordham M, Lovekamp WE, Thomas DSK, Phillips BD (2013) Understanding social vulnerability. In: Thomas DSK, Phillips BD, Lovekamp WE, Fothergill A (eds) Social vulnerability to disasters. CRC Press, Boca Raton, London, New York, pp 1–29

Fox-Rogers L, Devitt C, O'Neill E, Brereton F, Clinch JP (2016) Is there really "nothing you can do"? Pathways to enhanced flood-risk preparedness. J Hydrol 543(B):330–343

Głosińska E (2014) Spatial planning in floodplains for implementation by the floods directive in Poland. Geogr Pol 87(1):127–142

Gradlyan A, Baghdasaryan B (2013) YouTube videos as a new source for content analysis. Caucasus Research Resource Centers, Tbilisi

Gvoždíková B, Müller M (2017) Evaluation of extensive floods in western/central Europe. Hydrol Earth Syst Sci 21(7):3715–3725

Halama A (2013a) Polityka przestrzenna na terenach zalewowych w małych miastach. Stud Ekon 144:311–322

Halama A (2013b) Zrównoważony rozwój małych miast w aspekcie zagrożenia powodziowego. Acta Univ Lodziensis Folia Geogr Socio-Oeconom 15:255–269

Henstra D, Thistlethwaite J, Brown C, Scott D (2018) Flood risk management and shared responsibility: exploring Canadian public attitudes and expectations. J Flood Risk Manag. https://doi.org/10.1111/jfr3.12346

Hillhorst D, Bankoff G (2004) Introduction: mapping vulnerability. In: Bankoff G, Freaks G, Hillhorst D (eds) Mapping vulnerability: disasters, development & people. Earthscan, London-Sterling, VA

Höppner C, Bründl M, Buchecker M (2010) Risk communication and natural hazards. CapHaz-Net WP5 report. Swiss Federal Research Institute WSL. http://caphaznet.org/outcomes-results/CapHaz-Net_WP5_RiskCommunication.pdf (25.02.2017)

Jemielniak D (2013) Netnografia, czyli etnografia wirtualna—nowa forma badań etnograficznych. Prakseologia 154:97–116

Jina YH, Junghyun K (2012) Obesity in the new media: a content analysis of obesity videos on YouTube. Health Commun 27:86–97

Konieczny R, Kundzewicz ZW, Siudak M, Działek J, Biernacki W (2017) Education and information as a basis for flood risk management—practical issues. In: Tyszka T, Zielonka P (eds) Large risk with low probabilities: perceptions and willingness to take preventive measures against flooding. IWA Publishing, London, pp 177–201

Kowalewski Z (2006) Powodzie w Polsce—rodzaje, występowanie oraz system ochrony przed ich skutkami. Woda–Środowisko–Obszary Wiejskie 6(1):207–220

Kuhlicke C, Scolobig A, Tapsell S, Steinführer A, DeMarchi B (2011) Contextualizing social vulnerability: findings from case studies across Europe. Nat Hazards 58(2):789–810

Kundzewicz ZW, Zalewski M, Kędziora A, Pierzgalski E (2010) Zagrożenia związane z wodą. Nauka 4:87–96

Lindell MK (2013) Risk perception and communication. In: Bobrowsky PT (ed) Encyclopedia of natural hazards, Springer, Dordrecht, pp 870–874

Luo C, Zheng X, Zeng DD, Leischow S (2014) Portrayal of electronic cigarettes on YouTube. BMC Public Health 14. http://www.biomedcentral.com/1471-2458/14/1028 (26.03.2017)

Magnuszewski A, Porczek M (2015) Wskaźnik potencjału powodziowego i względna ekspozycja na niebezpieczeństwo powodziowe gmin w Polsce. Prace Stud Geogr 57:55–65

McMillan S (2000) The microscope and the moving target: the challenge of applying content analysis to the World Wide Web. Journal Mass Commun Q 77(1):80–98

Mees H, Crabbé A, Alexander M, Kaufmann M, Bruzzone S, Lévy L, Lewandowski J (2016) Coproducing flood risk management through citizen involvement: insights from cross-country comparison in Europe. Ecol Soc 21(3). http://dx.doi.org/10.5751/ES-08500-210307

Mercer J, Kelman I, Lloyd K, Suchet-Pearson S (2008) Reflections on use of participatory research for disaster risk reduction. Area 40(2):172–183

Nasim A, Blank MD, Cobb CO, Berry BM, Kennedy MG, Eissenberg T (2014) How to freak a Black & Mild: a multi-study analysis of YouTube videos illustrating cigar product modification. Health Educ Res 29(1):41–57

Neuvel JMM, van der Knaap W (2010) A spatial planning perspective for measures concerning flood risk management. Int J Water Resour Dev 26(2):283–296

Powódź. W obliczu zagrożenia (2013) Rządowe Centrum Bezpieczeństwa, Wydział Analiz RCB, Warszawa

Raška P (2015) Flood risk perception in Central-Eastern European members states of the EU: a review. Nat Hazards 79(3):2163–2179

Roth F, Brönnimann G (2013) Focal report 8: risk analysis using the internet for public risk communication. Risk and Resilience Research Group, Center for Security Studies (CSS), ETH Zürich

Rufat S, Tate E, Burton CG, Maroof AS (2015) Social vulnerability to floods: review of case studies and implications for measurement. Int J Disaster Risk Reduct 14:470–486

Spence PR, Lachlan KA (2010) Disasters, crises, and unique populations: suggestions for survey research. In: Ritchie LA, MacDonald W (eds) Enhancing disaster and emergency preparedness, response, and recovery through evaluation. New directions for evaluation, 126, pp 95–106

Steinführer A, Kuhlicke C (2007) Social vulnerability and the 2002 flood: country report Germany (Mulde River). FLOODsite report T11-07-08

Steinführer A, Kuhlicke C, De Marchi B, Scolobig A, Tapsell S, Tunstall S (2009) Local communities at risk from flooding: social vulnerability, resilience and recommendations for flood risk management in Europe. FLOODsite, Leipzig

Tapsell S, McCarthy S, Faulkner H, Alexander M (2010) Social vulnerability and natural hazards. CapHaz-Net WP4 report. Flood Hazard Research Centre—FHRC, Middlesex Univer-

sity, London. http://caphaz-net.org/outcomes-results/CapHaz-Net_WP4_Social-Vulnerability2.
    pdf (15.05.2016)
Tate E (2012) Social vulnerability indices: a comparative assessment using uncertainty and sensi-
    tivity analysis. Nat Hazards 63:325–347
Thomas DSK, Phillips BD, Lovekamp WE, Fothergill A (eds) (2013) Social vulnerability to disas-
    ters. CRC Press, Boca Raton, London, New York
Understanding the risks, empowering communities, building resilience: the national flood and
    coastal erosion risk management strategy for England (2011) Environment Agency, Bristol
Yoon DK (2012) Assessment of social vulnerability to natural disasters: a comparative study. Nat
    Hazards 63:823–843
*Zagrożenia okresowo występujące w Polsce* (2013) Rządowe Centrum Bezpieczeństwa, Wydział
    Analiz RCB, Warszawa
Żelaziński J (2011) Nauczmy się żyć z powodziami. Let's learn to live with flooding. Infos, Biuro
    Analiz Sejmowych 2:1–4

# Chapter 2
# The Course of Floods in the Study Areas and Their Consequences

Nearly every year, southern Poland experiences raised river water levels, and years with several such occurrences within a single season are not rare. Against the background of Poland as a whole, southern Poland is exposed to the greatest flood risk (Maciejewski 2000). The rivers in the Carpathian Mountains and the Sudety Mountains are characterised by the highest flood potential index (Stachý et al. 1996; Bartnik and Jokiel 2007, 2008, 2012). In addition, towns located, in particular, along the Vistula River and its tributaries within the Sandomierz Basin are characterised by a high level of housing development in the flood hazard zone. Locally, such areas account for more than 20% of the areas of municipalities (Magnuszewski and Porczek 2015).

This chapter presents the meteorological and hydrological characteristics of the floods which have affected southern Poland over the last 30 years and for which field studies have been carried out. Case studies are represented by localities that were flooded during major regional floods in 1997, 2001 and 2010. Most of the flood events studied came from the 2010 flood, which actually consisted of three flood waves—in May, June and August. Local flash floods from 2009, 2012 and 2014 were also chosen representing more isolated flood events.

## 2.1 Flood in July 1997

In early July 1997, shallow low-pressure systems were lying over nearly all of Europe, with centres over the UK, the Norwegian Sea and Finland. In parallel, the Azores High developed above the Atlantic. On 3 July, a shallow low with weather fronts started to arrive in Poland to turn into a quasi-stationary system the next day. It led to the separation of the air masses lying over Poland which differed in terms of air mass humidity, causing gradual build-up of clouds along the front and triggering precipitation. North-eastern Poland was under the influence of hot tropical air masses, with cool and very humid polar and sea air inflowing over the rest of the country (Olszowicz et al. 1999).

© The Author(s), under exclusive licence to Springer Nature Switzerland AG 2019
J. Działek et al., *Understanding Flood Preparedness*, SpringerBriefs in Geography,
https://doi.org/10.1007/978-3-030-04594-4_2

Vertically developed nimbostrati with a very high moisture content started to build up over southern Poland. Clouds with the greatest thickness and moisture content were recorded over the upper Odra river basin, where the compact cloud layer reached 8–9 km. This resulted in continuous heavy rainfall, which was additionally intensified by the development of storms with showers and by precipitation caused by the pressure of air masses pressing against the barrier formed by the Sudetes (Olszowicz et al. 1999).

Heavy rainfall started to appear in the upper Odra basin as early as 4 July (Dubicki and Malinowska-Małek 1999). The precipitation which was directly responsible for the flood in question began on 5 July in the evening and lasted from 50 h in the Nysa Kłodzka catchment area to 70 h in the Biała Lądecka basin. Two centres with the highest precipitation developed within the Odra river basin. One was located in the eastern part of the Kłodzko Basin, and the other within the Biała Głuchołaska river basin. In the eastern part of the Kłodzko Basin, within the Biała Lądecka and Wilczka basin, precipitation ranged from 316.2 mm in Bolesławów to 482.2 mm in Kamienica (Fig. 2.1). The amount of precipitation east of Jesenik (in the Biała Głuchołaska catchment area) totalled 512.0 mm. At the same time, it should be noted that most of the precipitation measured at the time was recorded over only 3 days (5–7 July, Dubicki and Malinowska-Małek 1999).

The highest daily precipitation was recorded on 5 July, when it ranged, in the eastern part of the Kłodzko Basin, from 94.8 mm in Lądek Zdrój to 200.1 mm in Międzygórze. Between 5 and 9 July, the precipitation in the area totalled from 150 to 250% of typical rainfall recorded there in July. Within the area record-

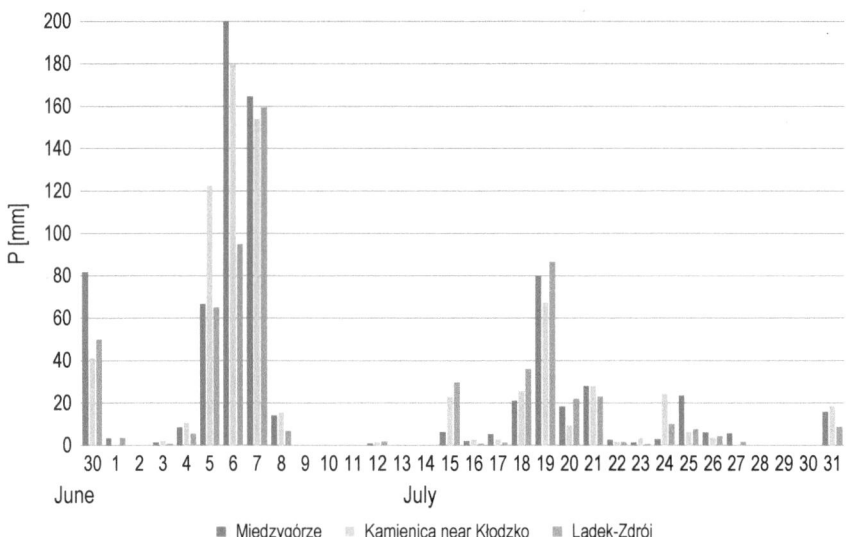

**Fig. 2.1**  Daily precipitation in the Kłodzko Basin between 30 June and 31 July 1997. *Source* Own elaboration using data from the IMGW-PIB

ing the highest rainfall, this was even more than 300% (with the maximum in Międzygórze—346.2%) (Dubicki and Malinowska-Małek 1999).

The July 1997 flood in the Odra river basin caused such huge financial damage and affected society to such a great extent that it was called the 'Millennium Flood'. It developed most rapidly within the Kłodzko Basin, where, as a result of the very high sloping and high rainfall intensity, water started to rise almost immediately after it had started to rain. The time from the occurrence of the maximum intensity of precipitation to the culmination of the flood wave on individual rivers was from just 3 h on the Biała Lądecka River in Lądek-Zdrój to 9 h on the Nysa Kłodzka in Kłodzko (Dubicki 1999).

Flood waves on the Kłodzko Basin rivers began to form in the morning of 6 July. A rapid rise of water levels, especially of the right tributaries of the Nysa Kłodzka: the Biała Lądecka and Wilczka Rivers, as well as the Nysa Kłodzka itself in its upper reaches, was seen in the afternoon of 7 July. The culmination of the flood wave occurred between 20.00 and 22.00 h (Dubicki 1999), and was 1340 $m^3$ $s^{-1}$ on the Nysa Kłodzka in Kłodzko. A key role in the development of such a high discharge on the Nysa Kłodzka was played by the Biała Lądecka River, with a discharge of 700 $m^3$ $s^{-1}$ at its mouth in Żelazno. However, the small Wilczka River rose the most, recording a maximum discharge within its 35 $km^2$ catchment area of 150 $m^3$ $s^{-1}$. The specific runoff from the catchment area amounted to 4250 $dm^3$ $s^{-1}$ $km^{-2}$, while that from the Biała Lądecka basin was 2560 $dm^3$ $s^{-1}$ $km^{-2}$ in Lądek Zdrój (Fig. 2.2; Dubicki 1999). The maximum discharge of the Biała Lądecka was at 795% (in Lądek Zdrój) and 961% (in Żelazno) of the mean of maximum discharges (MMD) for the period 1981–2010.

The economic losses caused by the catastrophic 1997 flood across the country were estimated at PLN 12.8 billion (Egler 2003; Kundzewicz and Matczak 2010),

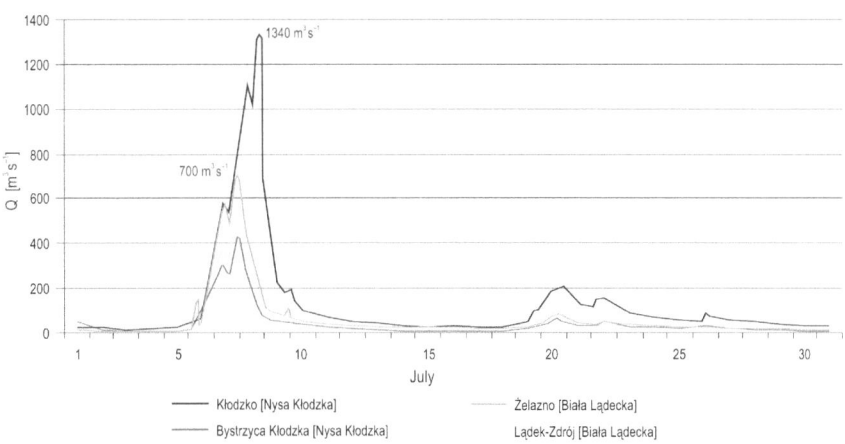

**Fig. 2.2**  Hydrographs for rivers in the Kłodzko Basin, 1–31 July 1997. *Source* Own elaboration using data from the IMGW-PIB

72% of which was recorded in the Odra river basin. The highest losses were recorded in what was then the Wałbrzyskie Region, where they amounted to PLN 1.34 billion (*Dorzecze* 1999; Dubicki et al. 1999a, b). Around 10 thousand buildings were damaged or completely destroyed in that region, and the associated losses were estimated at PLN 181.3 million. In addition, 1677 km of roads and 187 bridges were destroyed in the region, with the related damage totalling PLN 704.7 million. The total number of deaths across the country was 55 (Dubicki et al. 1999a, b).

## 2.2  Flood in July 2001

The July 2001 flood in the upper Skawa basin was caused by rapid rainfall which resulted from a deep low from the Hungarian Plain approaching Poland. After crossing the arc of the Carpathians, it stalled for a few days over south-east Poland (Lach 2002; Lach and Lewik 2002). The pressure situation which developed at the time over central Europe drove warm air masses northwards, and cool and humid masses in the opposite direction (Franczak 2013; Franczak and Listwan 2015). After encountering the orographic barrier formed by the Carpathians, the moving air masses formed a thick layer of clouds, triggering intense rainfall, including violent thunderstorms. During the storms, precipitation highly intensified.

In the first decade of July, from 37 to 78 mm of rain fell onto the area under study, causing strong hydration of the catchment area. On 16–17 July, the area saw another 30–65 mm (Fig. 2.3; Franczak and Listwan 2015). On 19–20 July, rainfall intensified within the catchment area of the Skawica, and from 22 July onwards, it extended to the entire basin of the Skawa. Within 2 days (22–23 July), it amounted to 45–73.5 mm, reaching the maximum intensity between 24 and 27 July (Franczak 2013).

On 24 July, rainfall within nearly the entire catchment area of the Skawa (except for the Raba Valley) ranged from 61.9 mm in Zawoja to 94.6 mm in Maków Podhalański. On the next day, its intensity declined nearly across the basin, but the central part saw catastrophic rainfall of up to 190.8 mm in Maków Podhalański (Franczak 2013; Pociask-Karteczka and Żychowski 2014; Franczak and Listwan 2015). The heaviest rainfall occurred over Makowska Góra Mountain between 15.00 and 18.00 h, when precipitation of approximately 150 mm was recorded (Bryndal 2014). However, the extent of the rainfall was limited since the station in Bogdanówka, which lies 7 km away from Makowska Góra Mountain, recorded rainfall of 68.7 mm. On the next day, rainfall in the Skawa basin reached 64 mm and stopped completely on 28 July (Franczak 2013).

During the 13 days of continuous rainfall (15–27 July), precipitation within the Skawa basin ranged from 243 to 316 mm, amounting to 457.8 mm in Maków Podhalański. During the 5 days with the heaviest rain, the Maków Podhalański station recorded precipitation of 392.7 mm (Franczak 2013).

The July 2001 flood in the Skawa basin had been the greatest flood since the start of measurements by the Sucha Beskidzka station in 1886 (Franczak and Listwan 2015).

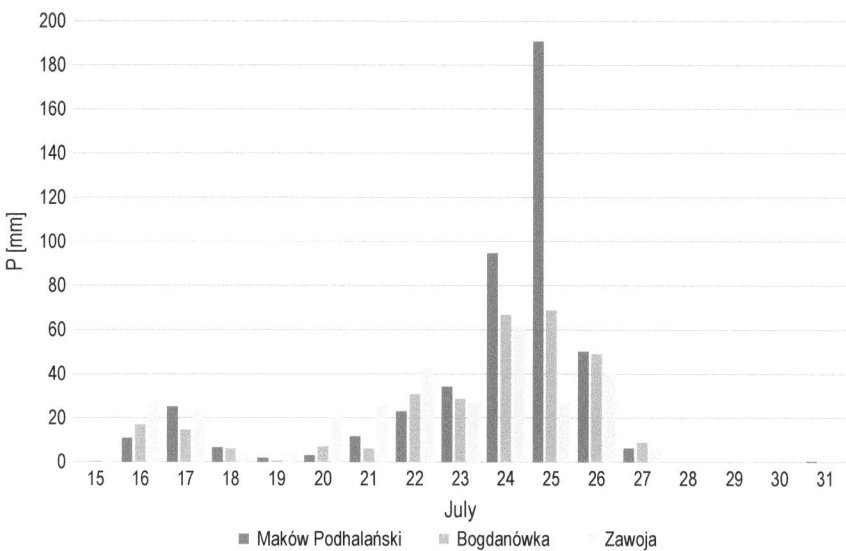

**Fig. 2.3** Daily precipitation in the upper Skawa basin between 15 and 31 July 2001. *Source* Own elaboration using data from the IMGW-PIB

The water level started to rise rapidly on 24 July. On the next morning, the discharge of the Skawa in Sucha Beskidzka increased to 308 $m^3$ $s^{-1}$ to reach the maximum value of 737 $m^3$ $s^{-1}$ at 21.00 h (Fig. 2.4; Franczak 2013). Within the catchment area of the Paleczka, which is not monitored and where the floods caused the greatest damage, the culmination of the flood wave was observed between 20.00 and 21.00 h (Tomica and Żuławska 2002). The stream gauge at the newly built Świnna Poręba dam on the Skawa (below the mouth of the Paleczka) recorded a maximum discharge of 1019 $m^3$ $s^{-1}$ (Franczak 2013).

During the flood, the specific runoff from the Skawa basin at the stream gauge in Osielec amounted to 1959 $dm^3$ $s^{-1}$ $km^{-2}$ and was one of the highest among the Polish rivers with similar catchment areas. The maximum discharge of the Skawa was 377% (in Sucha Beskidzka) and 365% (in Wadowice) of the mean of maximum discharges (MMD) for 1981–2010.

The estimated losses caused by the 2001 flood within the entire Vistula catchment basin were PLN 4.2 billion (*Wdrażanie dyrektywy* … 2013), of which a quarter (PLN 1.013 billion) was suffered by the Małopolskie Region. The damage to infrastructure in the Sucha Beskidzka County amounted to PLN 62.6 million, which was the second highest—after the Nowy Sącz County—value among all counties in the region. The Sucha Beskidzka County saw the highest rate of losses relative to its budget (*Powódź w wymiarze* … 2001). The losses in the Budzów municipality, which was the most seriously affected municipality of all, amounted to PLN 24 million (*Program Odbudowa* … 2003).

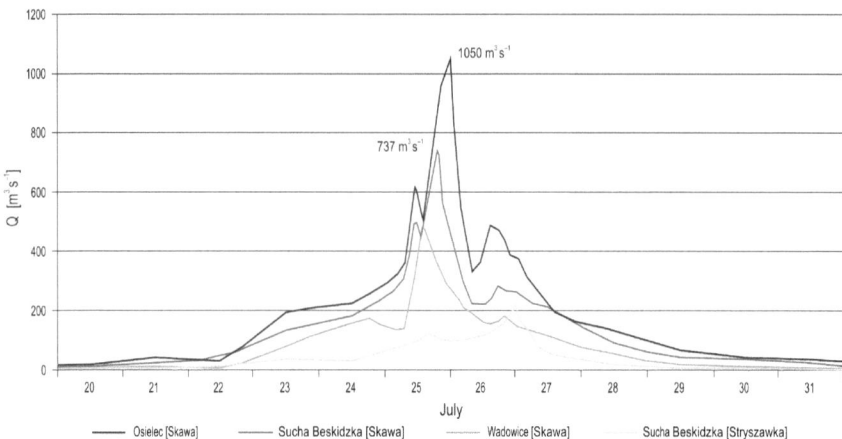

**Fig. 2.4** Hydrographs for the Skawa and the Stryszawka Rivers, 20–31 July 2001. *Source* Own elaboration using data from the IMGW-PIB

As a result of the floods, 761 buildings, 8 km of national roads, more than 20 km of regional roads, 59 km of county roads and 277 km of municipal roads were destroyed or damaged. In addition, 11 hydrotechnical structures were destroyed, and damage was recorded over an area of 302 ha of arable land and 277 ha of grassland (Trybała and Przywarska 2004). As a result of the flood, one person was killed in the basin of the Paleczka River.

## 2.3    Flood in June 2009

In the last decade of June 2009, warm tropical air masses characterised by high convection energy levels started to arrive in Poland. An anticyclone started to build up over Scandinavia, which—as a result of blocking the movement of a cyclone developing in the Balkans—drove warm and humid air masses towards Poland. This led to the formation of a longitudinally warm weather front, with air temperature reaching 15 °C. Frontal clouds developed, inside which extensive convection cells were building up, with their tops reaching the tropopause (Olędzki 2009; Bryndal et al. 2010).

Precipitation in the Czarna basin and in its vicinity began on 19 June. Daily precipitation at the individual weather stations ranged from a dozen to several tens of millimetres, increasing the hydration of the catchment area. Higher rainfall was recorded in Pilzno, where it amounted to 39.1 mm on 23 June, and to 70.8 mm on 24 June (Fig. 2.5). The Czarna river basin and the neighbouring area saw the heaviest rainfall on 26 June, which was associated with the occurrence of thunderstorms along the front moving from the east.

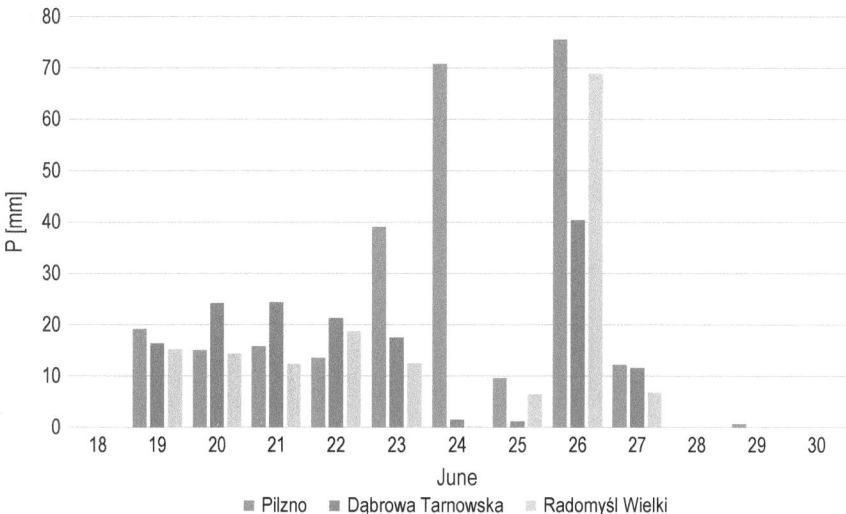

**Fig. 2.5** Daily precipitation in the Czarna basin and its vicinity between 18 and 30 June 2009. *Source* Own elaboration using data from the IMGW-PIB

The rainfall in the catchment area of the Czarna began around 19.00 and continued for approximately 6 h. During the first hour, precipitation in the southern part of the catchment area ranged from several to a dozen or so mm, but then the rainfall increased in intensity. Between 21.00 and 22.00, hourly rainfall over the towns of Róża and Borowa was about 45 mm. During the next hour (22.00–23.00), the precipitation in the Czarna basin increased to about 60 mm. The heaviest rain was observed between 22.30 and 23.30, when hourly precipitation in the Czarna basin ranged from 60 to 70 mm (Fig. 2.6). On 26 June, at some measuring sites neighbouring on the Czarna basin, daily precipitation amounted to 75.6 mm in Pilzno, 68.9 mm in Radomyśl Wielki and 68.5 mm in Tarnów.

The rainfall which was observed in the catchment of the Czarna from 19 June onwards increased the discharge of the river in Głowaczowa from 0.5 $m^3$ $s^{-1}$ to nearly 10 $m^3$ $s^{-1}$ (25 June). A surge in the water level of the Czarna and its tributaries began on the night between 26 and 27 June. During the torrential rains, small streams, including Potok Borowy, Potok Wsiowy and other streams flowing from the direction of Stare Jaworniki and Jaźwiny, overflowed. The discharge of the Czarna increased at 7.00 h to 55.8 $m^3$ $s^{-1}$, and at about 9.00 h, it overflew its banks. The flood wave reached the Czarna at 14.00 h, with a maximum discharge of 94.6 $m^3$ $s^{-1}$ (Fig. 2.7). The maximum discharge of the Czarna in Głowaczów represented 271% of the mean of maximum discharges (MMD) in the 1981–2010 period.

As the flood wave on the Czarna River moved on, the entire bottom of the valley was flooded, with the width of the overflow area reaching up to approximately 280 m. The streets in the centre of the village and the area around the school and the sports stadium were flooded. Along the other streets, residential buildings and farm

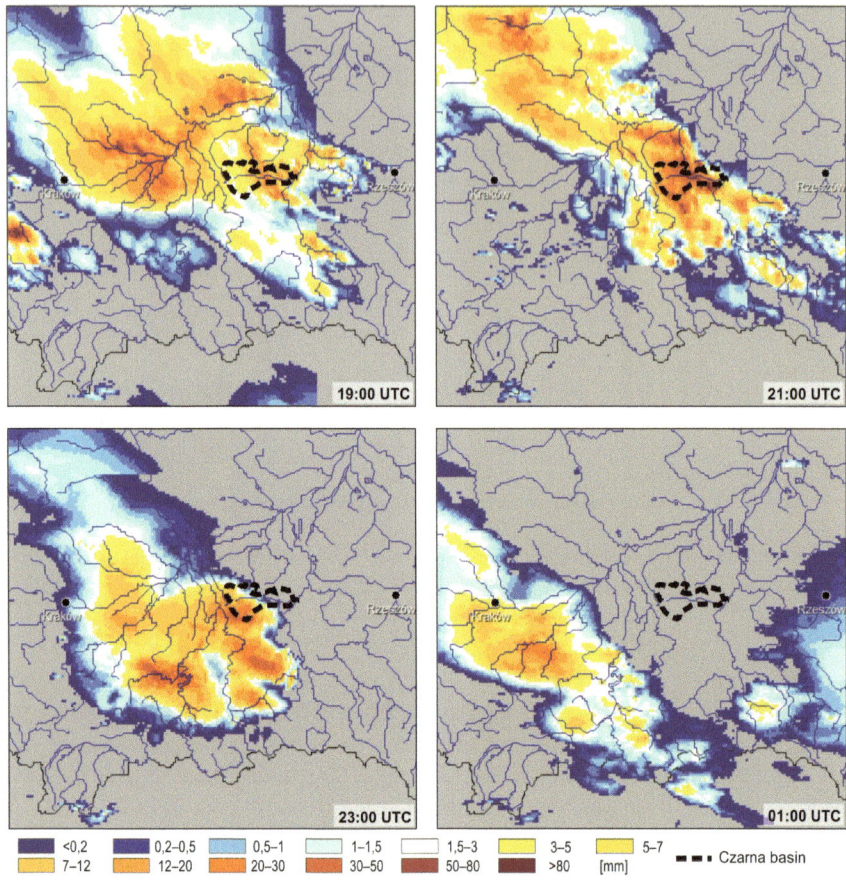

**Fig. 2.6** Distribution of hourly rainfall in south-eastern Poland on 26 June 2009 according to radar data. *Note* Based on hourly PAC rainfall data. *Source* Own elaboration using data from the IMGW-PIB

buildings located along the tributaries of the Czarna were flooded. Approximately, 60% of farms were flooded in the Borowa village (*Powódź w gminie Czarna* 2009).

## 2.4   Flood in May and June 2010

One of the largest floods in Poland in recent years was recorded in 2010, with as many as three flood waves forming in the catchment area of the upper Vistula (Nachlik and Kundzewicz 2016). The weather over Poland in May that year was determined by the Azores High, stationed over western Europe and by another anticyclone with the centre over Russia. In between, cyclones were moving across Poland from the

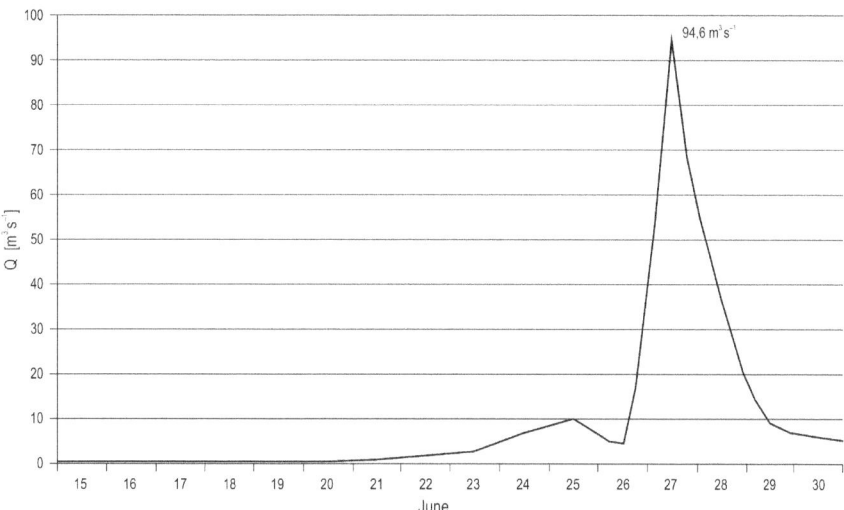

**Fig. 2.7**  Hydrograph for the Czarna River in Głowaczowa, 15–30 June 2009. *Source* Own elaboration using data from the IMGW-PIB

south of Europe towards Scandinavia (Cebulak et al. 2012; Zawiślak et al. 2012). In the first half of May (1–14 May), the cyclones produced rainfall in the upper Vistula basin ranging from about 100 to 150 mm, with the amounts rising locally to 180 mm (Cebulska et al. 2013; Śmiech 2012; Franczak 2013).

The first heavy rain was recorded in the upper Vistula basin on the night between 15 and 16 May, reaching the highest daily precipitation totals on 16 May. Almost across the whole Western Beskids, including the foothills, the daily rainfall totals exceeded 100 mm on that day. Two main precipitation centres formed over the Silesian Beskids and the Little Beskids, as well as in the Maków Beskids and the Island Beskids. The highest daily precipitation totals were recorded in Straconka (the Little Beskids)—185.2 mm (Fig. 2.8; Cebulak et al. 2012).

Very intense rainfall continued for the next 2 days (17–18 May). Four-day (May 15–18) rainfall ranged between 374.5 mm in Ustroń-Równica-Wieś (the Silesian Beskid) and 368.6 mm in Straconka, which also recorded the highest precipitation in May, amounting to 593.5 mm (Cebulak et al. 2012). During the main rainfall wave, the highest amounts of rain were observed in the Silesian Beskids and the Silesian Foothills, and slightly lower along a belt extending from those regions to the Island Beskids. High precipitation levels (>250 mm) were also recorded in the Tatras, and slightly lower in the Żywiec Beskids, as well as within the Podhale-Magura Area.

During the first wave of rainfall between 1 and 27 May, the precipitation ranged from 156 mm in Krempna and 226.8 mm in Jasło. In the same period, the amount of precipitation ranged from 213.5 mm in Tuchów to 343.9 mm in Ptaszkowa (Fig. 2.9).

The upper Vistula basin saw the second wave of the 2010 rainfall in early June, as a result of the build-up of a vertical low-pressure system consisting of two systems,

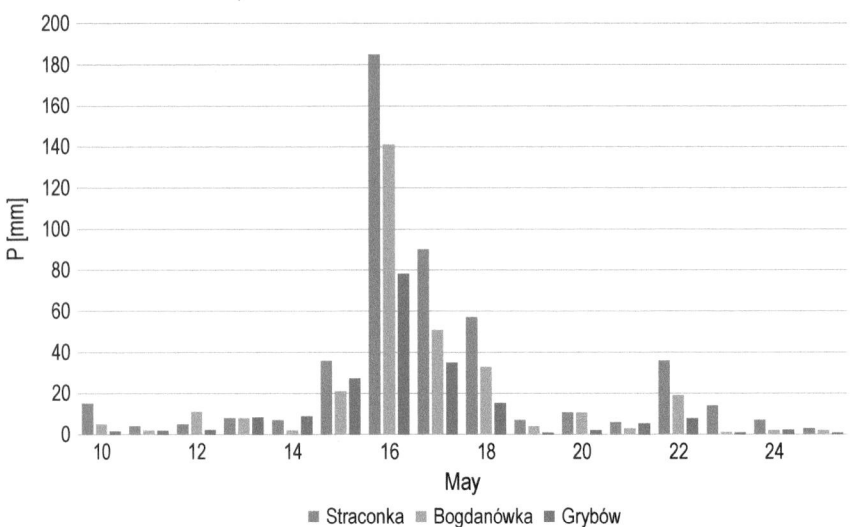

**Fig. 2.8**  Daily precipitation in the upper Vistula basin between 10 and 25 May 2010. *Source* Own elaboration using data from the IMGW-PIB

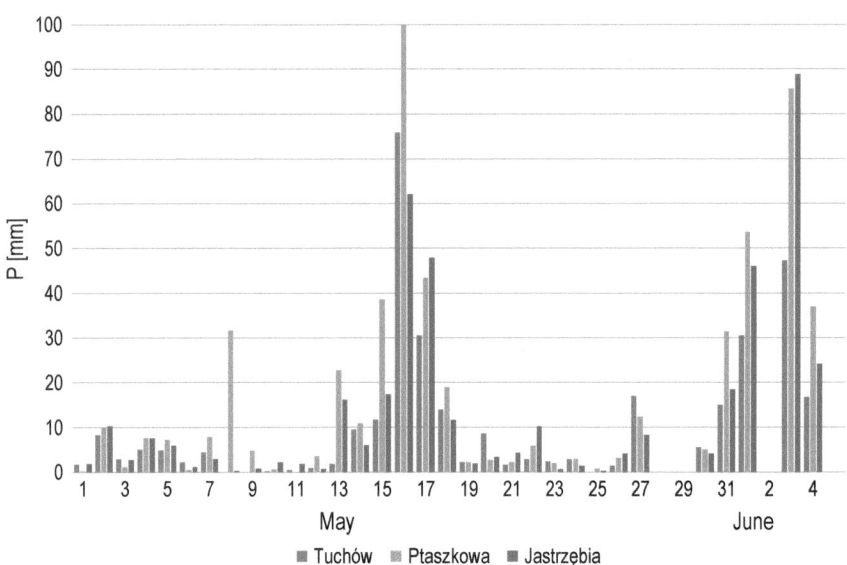

**Fig. 2.9**  Daily precipitation in the Biała Tarnowska basin between 1 May and 4 June 2010. *Source* Own elaboration using data from the IMGW-PIB

the first of which developed over Hungary and moved over Ukraine, while the other formed over Ukraine and moved to Poland. Their movement out of Poland was blocked by an anticyclone centred over the Norwegian Sea, with simultaneous influx of cold polar sea air from the west. This produced advection of warm and very humid air, leading to high rainfall, which was intensified by convection caused by the presence of orographic barriers. As a consequence, the rainfall lasted from 30 May to 4 June (Zawiślak et al. 2012).

During these 6 days with rain, two centres with the highest precipitation levels formed in the Carpathians: one over the Tatra Mountains, where rainfall ranged from 220 to 250 mm (maximum within the Hala Gąsienicowa mountain pasture—250.2 mm), and the other in the upper reaches of the Biała and Ropa basins. The heaviest rainfall was recorded at the station in Huta—218.0 mm, but it also exceeded 200 mm at several other measuring sites (Cebulak et al. 2012). Ptaszkowa recorded rainfall of 212.7 mm. On 3 June, the area saw the highest daily precipitation totals across the Carpathians. Within the Biała basin, it exceeded 80 mm and was higher than 100 mm locally. The greatest daily precipitation total was seen at the station in Huta, where it amounted to 124.1 mm, in Szymbark—107.0 mm (Wiejaczka and Bochenek 2013), and in Ptaszkowa, Jastrzębia and Wysowa, it ranged between 80 and 90 mm.

The heavy rainfall seen in the upper Vistula basin in May 2010 added to the formation of a flood wave on the Vistula and its tributaries. The flood wave on the Vistula River was supplied by two sources of floodwaters: the tributaries connecting the Mała Wisła and the Raba, the river basins of which saw high evenly spread rainfall and the other tributaries, where precipitation levels proved to be lower (Drezińska 2012b).

Heavy rain lasting from mid-May fell onto the already heavily soaked catchment area, which highly increased discharge in its watercourses (Fig. 2.10). The alert thresholds in the upper reaches of the Vistula and its tributaries were exceeded on 16 May, and as a result of further rapid rise in water levels, the maximum discharge levels were reached at most stream gauges on 17 and 18 May. In Krakow, the flood wave peaked on 18 May at 20.00 h, and within the Sandomierz Basin, the flood wave culmination on the Vistula was recorded between 19 and 21 May. The peak wave on the Vistula River reached Szczucin on 19 May at 10.00 h, and it arrived in Sandomierz on the same day, at 6.00 h. The differences between the times when the wave peaked along the individual sections were caused by numerous breaches in the flood banks and the lowering in peak wave height as a result of the propagation the floodwater behind the banks (Drezińska 2012b).

The 2010 flood in upper reaches of the Vistula and its Carpathian tributaries was among the largest in the history of measurements (*Raport … 2010*; Drezińska 2012a). As the catchment area grew bigger, the maximum discharge of the Vistula at the successive stream gauges increased from 429 $m^3$ $s^{-1}$ in Skoczów to 5400 $m^3$ $s^{-1}$ in Zawichost (Drezińska 2012a).

The May 2010 flood caused extensive damage in Bieruń. On the Vistula, the gauging stations in Jawiszowice and Nowy Bieruń recorded discharges indicative of 'millennial water'. The discharge of the Vistula in Nowy Bieruń amounted to 843 $m^3$ $s^{-1}$, and to 102 $m^3$ $s^{-1}$ on the Przemsza River (Drezińska 2012a). On 17 May, as a result of its runoff being hindered by the elevated waters of the Vistula, the waters of the Gostynia River rose, creating backwater, which resulted in a breach in the

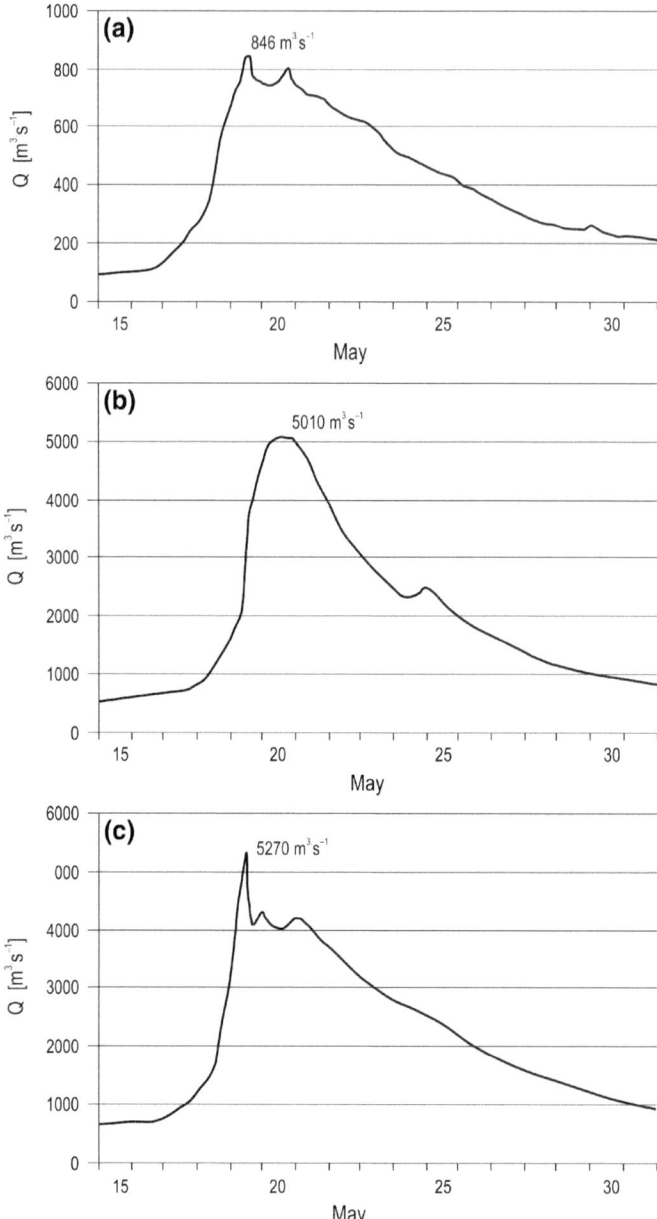

**Fig. 2.10** Hydrographs for the Vistula in Nowy Bieruń (**a**), Szczucin (**b**) and Sandomierz (**c**), 15–31 May 2010. *Source* Own elaboration using data from the IMGW-PIB and based on Maciejewski et al. (2011)

left-bank levee of the river at 22.30 h. The floodwaters flowing through the breach started to propagate across the landside of the levee, flooding the highly developed area within a former meander of the Vistula, in the districts of Kopań, Bijasowice and Zabrzeg. This created a vast reservoir since runoff of the water from the flooded area was hindered. On 21 May, the authorities decided to make a breach in the levee on the Vistula, which made it possible for the waters to be partly removed from the overflow area. Despite this, the water stagnated within lower areas for another 10 days.

Similarly to other reaches, the flood wave on the Vistula River in Szczucin was characterised by a very steep rising phase. The peak discharge was 5010 m$^3$ s$^{-1}$. In Szczucin, the water level on the Vistula remained high much longer than in Sandomierz, with near-peak states recorded for 11 h. The declining phase of the flood wave lasted 12 days (Drezińska 2012b).

An extensive overflow area was formed in Tarnobrzeg, as a result of which the flood wave in Sandomierz had a very steep rising phase and a relatively short time of near-peak levels (merely 4 h). After attaining a maximum discharge of 5270 m$^3$ s$^{-1}$, the wave subsided to 4200 m$^3$ s$^{-1}$.

Another flood wave formed in the upper Vistula basin at the turn of May and June. At the time, the rivers of the central and eastern parts of the Polish Carpathians swelled more than in May. The flood wave on the Vistula was much lower than in May. By contrast, within the Biała Tarnowska and Wisłoka basins, the absolute maxima of the flood waves were exceeded (Figs. 2.11 and 2.12). The wave demonstrated the fastest rise on the Biała Tarnowska, with its peaking in Koszyce on 4 June. The maximum discharge was 836 m$^3$ s$^{-1}$. The waters on the Wisłoka surged in a similar way, with the wave peaking at the Łabuzie gauging station on 5 June, and the maximum discharge amounting to 1370 m$^3$ s$^{-1}$. Across the catchment area of the Wisłoka River, the maximum discharges neared the likelihood of 'centennial water', with the exception of the Ropa basin, where the flood wave peak exceeded the likelihood of 'millennial water'.

The maximum discharge of the Biała Tarnowska represented from 233% (in Ciężkowice) to 546% (in Grybów) of the mean of maximum discharges (MMD) for 1981–2010. On the Wisłoka, the values were growing along with the increase

**Fig. 2.11** Hydrograph for the Wisłoka River in Krajowice, 1–10 June 2010. *Source* Own elaboration using data from the IMGW-PIB

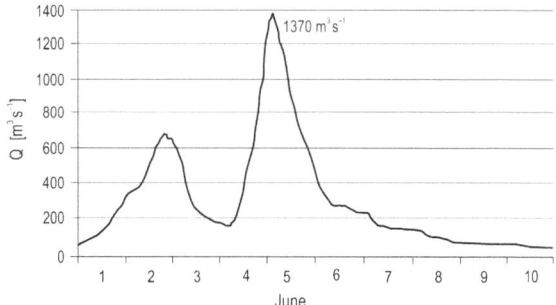

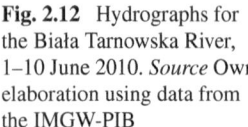

**Fig. 2.12** Hydrographs for the Biała Tarnowska River, 1–10 June 2010. *Source* Own elaboration using data from the IMGW-PIB

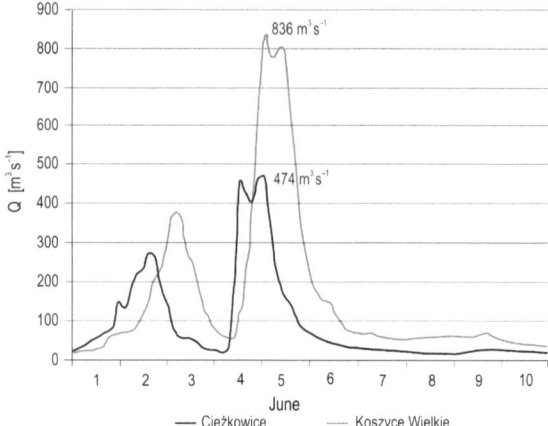

in the surface area of the catchment basin—from 156% (in Żułków) to 281% (in Łabuzie).

The May flood resulted in the inundation of Bieruń, where the overflow area reached 1500–1600 ha. The extent and scale of the flooding surprised many residents since they were unaware that some houses were located within flood risk areas, which had subsided several metres as a result of mining damage.[1] More than 350 residential buildings, 290 farm buildings and several public utility facilities were flooded. 371 households (730 persons) suffered damage, and 2300 people were evacuated during the disaster. Private property losses were estimated at PLN 30 million, damage to business property at PLN 10 million, and to municipal property at PLN 54.1 million (*Raport … 2011*; *Wystąpienie pokontrolne (MOPS w Bieruniu) 2011*).

Initially, the flood hazard in the Szczucin municipality was caused by small water-courses flowing across the municipality. The catastrophic flood started after levees were breached in the neighbouring Czermin municipality, after the waters of the Breń River and backwater from the Vistula River spilled towards Słupiec, Załuże, and then towards Dąbrowica, Maniów, Borki and Wola Szczucińska. Another flood wave on the Vistula River in June flooded these areas once again. After the runoff of stormwater into the Vistula was cutoff with sluice gates as a result of the high level of its waters, the towns of Laskówka, Delastowska, Lubasz and Łęka Szczucińska were flooded. The flooding affected 11.3 thousand ha of the municipality, i.e. about 94% of its area (*Wystąpienie pokontrolne (Urząd Miasta i Gminy Szczucin) 2013*). The floodwaters of the Vistula and the Breń in the Szczucin municipality spilled over a total of 7 thousand ha, mostly in the towns of Słupiec (1800 ha), Zabrnie (1500 ha), Maniów (1100 ha), Załuże (720 ha), Dąbrowica (570 ha) and Laskówka Delastowska (510 ha) (Jakóbik 2011).

In the Szczucin municipality, 1677 families were affected (*Sprawozdanie … b.d.*; *Wystąpienie pokontrolne (Urząd Miasta i Gminy Szczucin) 2013*). In the neighbour-

---

[1]Information obtained during interviews.

ing municipalities, the flood affected 220 households (680 people) in the Wadowice Górne municipality and 81 households (260 people) in the Czermin municipality (Lipińska 2011).

On 19 May 2010, after the levee on the Vistula River was broken in Koćmierzów, the following housing estates of Tarnobrzeg were flooded completely: Wielowieś, Sielec, Zakrzów, and partly Dzików and Sobów. Also, the right bank part of San-domierz was overwhelmed, and further downstream, following the breach of the flood bank on the Trześniówka River, a dozen villages in the neighbouring Gorzyce municipality were flooded. The waters of the Vistula and Trześniówka, which were spilling further east, stopped only on the right embankment of the Łęg River (Rybak 2011). In total, 2.5 thousand people left the flooded areas. In the area of Tarnobrzeg, the flood caused three deaths.

On 5 June, during the second wave of the flood, a sandbag ring levee built pre-viously around a breach in the flood bank was broken, as a result of which water of the Vistula re-entered the previously inundated areas (*Sprawozdanie … 2010; Stan ochrony … 2011; Wystąpienie pokontrolne (Urząd Miasta Tarnobrzega) 2015*). An area of 72 km² was flooded, with 36 km² of the land affected in Tarnobrzeg during the first wave, and 22 km² during the second wave. The depth of the flood-waters was 4 m, reaching 7 m at depressed spots (*Sprawozdanie … 2010; Rybak 2011*). The losses which affected public infrastructure amounted to PLN 43.6 mil-lion (*Wystąpienie pokontrolne (Urząd Miasta Tarnobrzega) 2015*). In Tarnobrzeg, about 1600 households and 5.2 thousand inhabitants were affected (Lipińska 2011).

In the Wisłoka basin, the earliest flood losses occurred during the May flood, but much greater damage was reported in early June. During the June flood, residential buildings in Brzostek and in many villages in the municipality were flooded. Damage was experienced by 245 households, affecting 912 people living in the residential buildings. The total losses in the municipality were estimated at PLN 18.2 million (*Powódź stulecia 2010; Wystąpienie pokontrolne (Urząd Miejski w Brzostku) 2013*). Thirty-seven households were affected in the neighbouring Pilzno municipality, with landslides causing much damage there (Lipińska 2011; *Wystąpienie pokontrolne (Urząd Miejski w Pilźnie) 2015*).

In the Tuchów municipality, the floodwater, which overtopped the levee of the Biała Tarnowska River creating a breach, flooded the town and several villages in the municipality. The overflow area formed in Tuchów covered 600 ha (Jakóbik 2011), affecting 94 families. Among the other towns of the Tuchów municipality, the greatest number of residential buildings was affected in the villages of Burzyn and Dąbrówka Tuchowska.

## 2.5  Flood in August 2010

In the first week of August 2010, the weather situation over Poland was determined by a low-pressure system moving north-east from the Mediterranean Sea. On 6 August, the weather front reached the borders between Poland, Germany and the

Czech Republic, where it came across the orographic barrier of the Jizera Mountains and the Lusatian Mountains. At the time, a front line associated with a tropospheric low-pressure system was formed, resulting in rainfall, which started on the morning of 6 June in the north of the Czech Republic, and lasted for 30–36 h. On 7 August, the front reached the catchment basin of the Miedzianka River and remained over the area for over 12 h (*Wspólny* ... 2010; Tokarczyk 2011).

On 6 August, rainfall started south of the Miedzianka basin, in the upper part of the Lusatian Neisse basin, and it also extended over a small fragment of the upper part of the Miedzianka basin. Bogatynia recorded 32 mm of rainfall, but the amount of precipitation was higher around the watershed area of the catchment basin. However, the amount of rainfall increased in the second half of the night from 6 to 7 August. Intensive precipitation was recorded between 2.00 and 3.00 h, when the Olivetska Góra station (south of the basin in question) recorded 46.8 mm of rainfall. At around 5.00, the first wave of rainfall, during which precipitation recorded at the Bogatynia station totalled 29 mm, stopped. At 7.00 h, the second wave of rainfall started, during which, by 11.00 h, the station adjacent to the Turów Coal Mine recorded 43 mm of rainfall (Salata et al. 2012), and the Bogatynia station—66 mm. Precipitation totals were much higher upstream of the catchment basin. The Hejnice and Mlýnice weather stations, which are located near the watershed, recorded 106 and 97 mm, respectively (Fig. 2.13). Between 8.00 and 9.00 h, the maximum rainfall intensity at the Hejnice station was 57.6 mm. At the same time, hourly precipitation in Bogatynia was 26.2 mm, and in the following hour, its intensity rose slightly to 27.6 mm.

On 7 August, Bogatynia recorded 160.2 mm of rainfall, and the Hejnice station, which neighbours on the upper part of the Miedzianka basin—179 mm. The Hejnice and Mlýnice stations recorded 2-day (6–7 August) precipitation totals of 252.4 and 250.2 mm, respectively, while the cumulative 2-day rainfall in Bogatynia was at a level of 192.2 mm. At the time, the highest rainfall in the upper part of the Lusatian

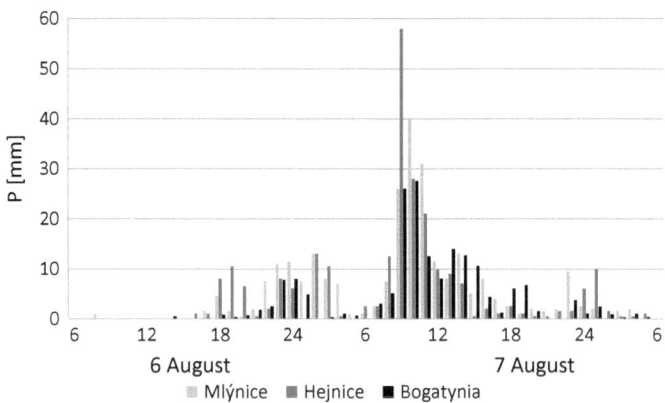

**Fig. 2.13** Hourly precipitation in the Miedzianka basin between 6 and 7 August 2010. *Source* Own elaboration using data from the IMGW-PIB

Neisse basin was recorded by the Olivetska Góra station (310.3 mm) (*Wspólny* …
2010; Tokarczyk 2011; Müller and Walther 2014).

In August 2010, the swollen waters of the Miedzianka triggered one of the greatest
floods in the history of Bogatynia. The flood wave started to build up on 6 August,
at 23.00 h. Over the night, the discharge of the Miedzianka River increased from
1.5 to 15.0 m³ s⁻¹. From 7.00 h, with the start of the main rainfall wave, the rate
of the rise in the water level grew, creating a peak flood wave within 5.5 h. At the
time, the discharge of the Miedzianka in Turoszów increased to 83 m³ s⁻¹ (Fig. 2.14;
Franczak and Listwan-Franczak 2016). The height of the flood wave on the mouth
section of the Miedzianka reduced as a result of a breach in the embankment along
its channel, through which floodwaters entered the areas of the nearby open brown
coal pit. The levee was broken at 13.30 h, i.e. when the level of the Miedzianka in
Turoszów reached a maximum. This prevented the peak wave from mounting further
downstream. Approximately, 3 million m³ of water penetrated the pit (Sondaj et al.
2011; Salata et al. 2012; Franczak and Listwan-Franczak 2016).

Upstream of the breach, the maximum discharge of the Miedzianka was approximately 190 m³ s⁻¹ (Fig. 2.14), which was slightly lower than discharge indicative of
the likelihood of 'millennial water'. The maximum specific runoff upstream of the
breach proved much higher than in Turoszów (1029 dm³ s⁻¹ km⁻²) and was 3050 dm³
s⁻¹ km⁻². The upper reaches of the catchment area, near the Polish–Czech border

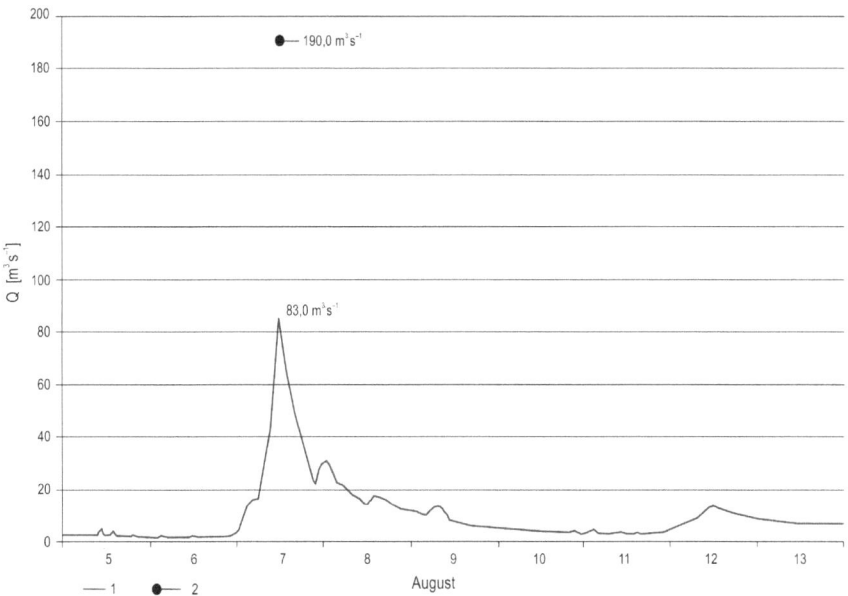

**Fig. 2.14** Hydrograph for the Miedzianka River in Turoszów, 5–13 August 2010. *Note* 1—hy-
drographic gauge on the Miedzianka in Turoszów; 2—maximum discharge of the Miedzianka as
estimated on the basis of traces of the great water above the flood bank breach. *Source* Own elaboration using data from the IMGW-PIB

in Markocice, the maximum specific runoff from the Miedzianka basin amounted to 4300 dm$^3$ s$^{-1}$ km$^{-2}$ (Franczak and Listwan-Franczak 2016). The maximum discharge of the Nysa Łużycka in Sieniawice represented 422% of the mean of maximum discharges (MMD) for the years 1981–2010.

During the flood in 2010, 11 buildings were destroyed completely, and overall 414 buildings were flooded and damaged, including 277 buildings and objects of historical value. During the flood, 600 persons were evacuated from the town and municipality of Bogatynia. There were also two deaths. In total, nearly 1200 families (approximately 3 thousand persons) were affected by the flooding across the municipality. The estimated losses associated with residential property totalled PLN 50 million, and the losses suffered by approximately 200 private business owners amounted to PLN 38.6 million (*Bogatynia po powodzi … 2011; Sprawozdanie dotyczące … b.d.; Wojewódzki raport … 2010; Powódź w Bogatyni … 2012*).

## 2.6   Flood in July 2012

On the first days of July 2012, two major pressure systems were developing in Europe. An extensive low-pressure centre was building up in south-eastern Europe, stretching from the Mediterranean Sea, across eastern Europe, northern Scandinavia, as far as the White Sea, and a low-pressure centre was forming over the north Atlantic, reaching with its mildly arched trough from the Jan Mayen island in the north, across the Norwegian Sea and the North Sea, France, as far as Gibraltar in the south. The resultant pressure system over Europe led to the formation of a cold weather front, which separated the hot air inflowing on its eastern side from the much cooler polar air on the opposite side. Hot air of tropical origin from the Mediterranean Sea and northern Africa was flowing from low towards high latitudes. A blocking anticyclone developed, which prevented zonal air circulation for the next several days. The front developed a ripple-like shape, creating small frontal waves within itself and sucking in hot air from the south and cool air from the north-west (Olędzki 2012a, b; Klejnowski 2012a, b).

The high thermal contrast and humidity inflowing from the eastern part of the Mediterranean See triggered convection phenomena in the frontal area. A squall line was formed at a distance of approximately 250–300 km from the front line, with much more intensive storms developing there than in the frontal zone itself. Such cell storms were developing on 5 July within the Sudety Mountains and the Sudety Foothills (Klejnowski 2012a, b). In the evening, two cell storms formed in the area of Świeradów-Zdrój and Kowary, and at approximately 14.40 h, they caused intensive rainfall within the upper Kaczawa basin. At around 17.40 h, the two cell storms joined together and were moving slowly northwards (to disappear between Złotoryja and Legnica). Rainfall within the upper Kaczawa basin continued until 19.50 h, following which a front reached the basin from the south-east, on which heavy rain was present from approximately 21.00 to 22.00 h.

On 5 July, precipitation in Kaczorów amounted to 59.2 mm, and in Świerzawa to 48.0 mm (Fig. 2.15), with much lower amounts of rainfall at the other stations adjacent to the catchment area under study (22.7 mm in Ciechanowice and 5.1 mm in Bolków). The highest precipitation was recorded along a 25-km-long and 10–12-km-wide belt stretching from the village of Muchów to Jelenia Góra (Fig. 2.16). The daily rainfall totals across the area amounted to 60 mm, reaching 72 mm in its northern part. Daily rainfall of 80–90 mm (max. 94 mm) was observed along a narrow strip which was 8 km in length and 3 km in width, running between the hills stretching from Skąpiec Mountain near Wojcieszów to the village of Janowice Wielkie in the Bóbr River valley.

At the turn of June and July 2012, the discharge of the Kaczawa River in the town of Świerzawa was approximately 0.7 m³ s⁻¹. Following a rise in the discharge on the Kaczawa to 19.7 m³ s⁻¹ on 5 July at 7.00 h, the Kaczawa and its tributaries saw a rapid surge in the evening. The flood wave on the Kaczawa peaked at 115 m³ s⁻¹ (Fig. 2.17). However, the greatest rise was recorded by the Bełczek and Olszanka Rivers (tributaries of the Kaczawa), whose floodwaters inundated entire valley bottoms. The flood wave descending from the upper reaches of the catchment basin filled the concrete-paved channel nearly completely, despite the fact that a large proportion of the flood wave was retained by a dry reservoir upstream of Wojcieszów. The city centre was flooded with swollen waters of the Olszanka stream. After coming together, the flood waves of the Kaczawa and its tributaries led to the creation within the valley bottom in Wojcieszów Dolny of a relatively large overflow area, which was approximately 150 m wide.

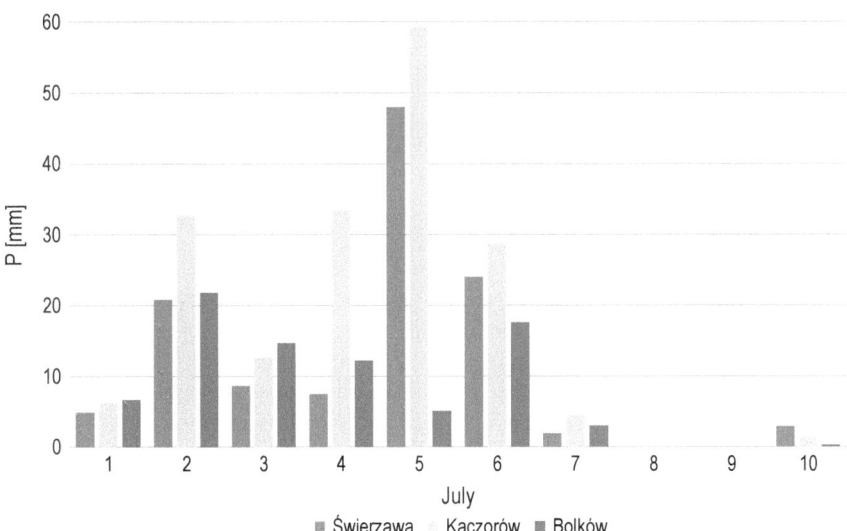

**Fig. 2.15** Daily precipitation in the upper Kaczawa basin between 1 and 10 July 2012. *Source* Own elaboration using data from the IMGW-PIB

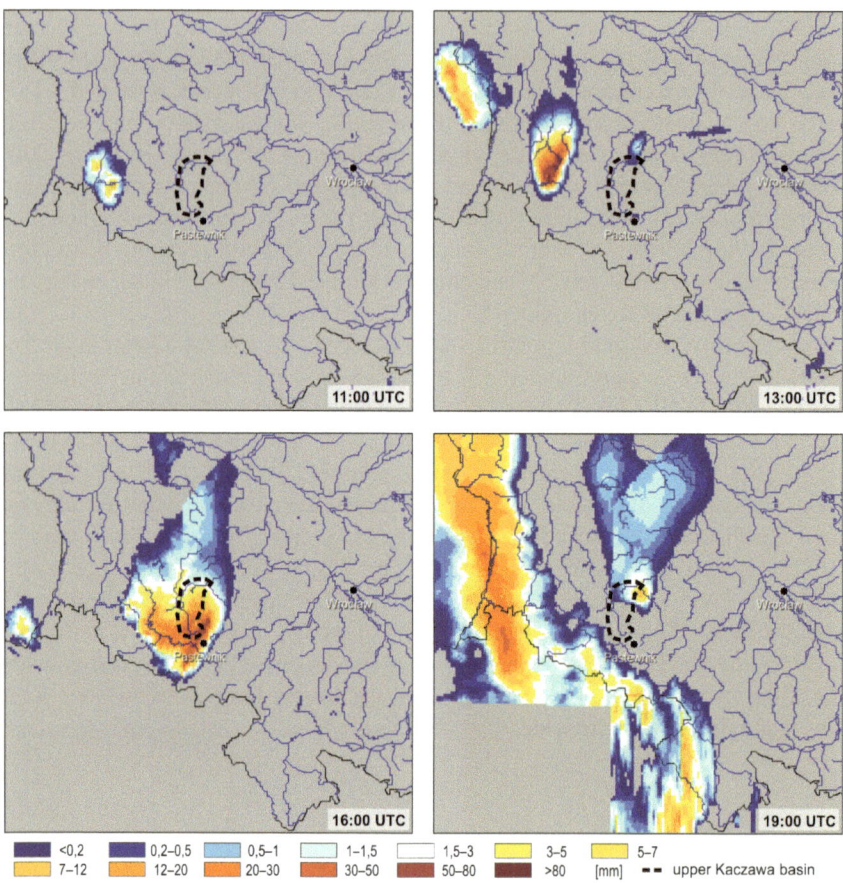

**Fig. 2.16** Distribution of hourly rainfall in South Western Poland on 5 July 2012 according to radar data. *Note* Based on hourly PAC rainfall data. *Source* Own elaboration using data from the IMGW-PIB

During the flood, the greatest damage was done by swollen waters of streams, which flooded residential buildings (including multi-family ones) and a school building, destroying sport and recreational facilities, too. Buildings located along the channel of the Kaczawa were flooded, as well.[2] After the flood, aid of PLN 325 thousand was paid within the municipality in 2012. The government issued a PLN 1.2 million promissory note to the local authorities for them to recover from the flood damage by repairing the municipal roads destroyed within the area of Wojcieszów (*Promesa … 2015*).

---

[2]Oral accounts by the residents of Wojcieszów and audiovisual coverage of the flood.

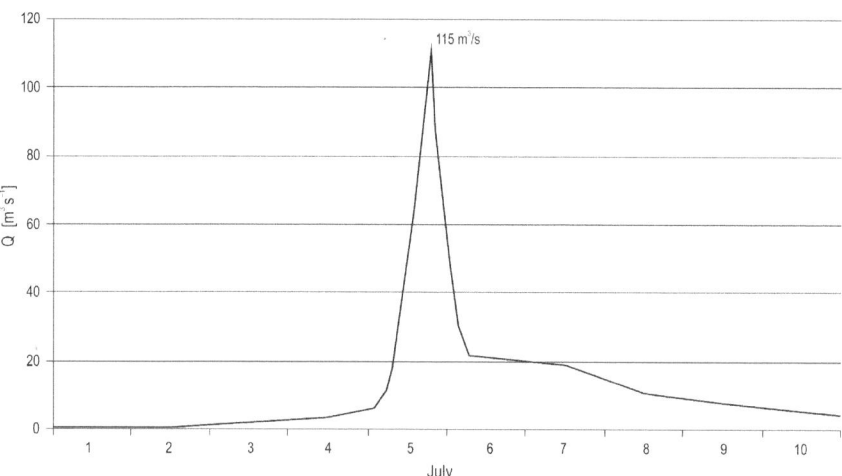

**Fig. 2.17** Hydrograph for the Kaczawa River in Świerzawa, 1–10 July 2012. *Source* Own elaboration using data from the IMGW-PIB

## 2.7 Floods in 2014

In the second decade of May 2014, at the northern foothills of the Babia Góra Massif, a flash flood occurred after heavy rainfall in the western part of the Carpathians. On 16 May, 138.0 mm of precipitation fell in Zawoja, and the 2-day sums of precipitation exceeded 200 mm. This resulted in a rapid flood in small forested catchments, from which the maximum unit outflow was about 2.2 $m^3$ $s^{-1}$ $km^{-2}$. The largest losses occurred in the Rybny Potok basin (area of 9.2 $km^2$) in Zawoja municipality (Franczak 2015).

Later that year, at the beginning of August 2014, after a dozen or so days of hot weather, a cool atmospheric front moved across Poland from the west with violent thunderstorms. On August 5, intense six hours of rainfall occurred in the Raba basin, and their total in Węglówka amounted to 95.2 mm. The highest intensity of precipitation was 31 mm/h. These precipitations caused an immediate flash flood in the Kasinianka catchment (48 $km^2$), at the mouth of which the maximum flow was 171 $m^3$ $s^{-1}$. The maximum unit outflow was 4.8 $m^3$ $s^{-1}$ $km^{-2}$, and the probability of its occurrence was 0.1% (so-called thousand-year water). The flood wave covered the entire bottom of the valley, causing the largest material losses in Kasinka Mała, located in Mszana Dolna municipality (Bryndal et al. 2017).

# References

Bartnik A, Jokiel P (2007) Odpływy maksymalne i indeksy powodziowości rzek europejskich. Gospodarka Wodna 1:28–32

Bartnik A, Jokiel P (2008) Odpływy maksymalne i indeksy powodziowości rzek półkuli północnej. Prz Geogr 80(3):343–365

Bartnik A, Jokiel P (2012) Indeksy powodziowości (Françou-Rodiera) i indeksy wysokiej wody w Karpatach i na Nizinach, w przekroju wieloletnim. Gospodarka Wodna 5:204–208

Bogatynia po powodzi. Liczby i zestawienia (2011) Niepublikowany materiał udostępniony przez Urząd Miasta i Gminy w Bogatyni

Bryndal T (2014) Powodzie błyskawiczne w małych zlewniach karpackich—wybrane aspekty zarządzania ryzykiem powodziowym. Ann Univ Paedagog Crac Stud Geogr 7(170):69–80

Bryndal T, Cabaj W, Gębica P, Kroczak R (2010) Gwałtowne wezbrania spowodowane nawalnymi opadami deszczu w zlewni potoku Wątok (Pogórze Ciężkowickie). In: Ciupa T, Suligowski R (eds) Woda w badaniach geograficznych. Instytut Geografii Uniwersytetu Jana Kochanowskiego, Kielce, pp 307–319

Bryndal T, Franczak P, Kroczak R, Cabaj W, Kołodziej A (2017) The impact of extreme rainfall and flash floods on the flood risk management process and geomorphological changes in small Carpathian catchments: a case study of the Kasiniczanka river (Outer Carpathians, Poland). Nat Hazards 88(1):95–120

Cebulak E, Kilar P, Milanówka D, Mizera M, Pryc R (2012) Wysokość, natężenie i przestrzenny rozkład opadów atmosferycznych. In: Maciejewski M, Ostojski MS, Walczykiewicz T (eds) Dorzecze Wisły: monografia powodzi maj–czerwiec 2010. IMGW-BIP, Warszawa, pp 27–41

Cebulska M, Szczepanek R, Twardosz R (2013) Rozkład przestrzenny opadów atmosferycznych w dorzeczu górnej Wisły. Opady średnie roczne (1952–1981). PK w Krakowie, IGiGP UJ, Kraków

Dorzecze Odry. Powódź 1997 (1999) Międzynarodowa Komisja Ochrony Odry przed Zanieczyszczeniem, Wrocław. www.mkoo.pl/download.php?fid=3918&lang=PL (5.10.2016)

Drezińska B (2012a) Przebieg fali powodziowej na Górnej Wiśle i jej dopływach. In: Maciejewski M, Ostojski MS, Walczykiewicz T (eds) Dorzecze Wisły: monografia powodzi maj–czerwiec 2010. IMGW-BIP, Warszawa, pp 51–62

Drezińska B (2012b) Ocena rozmiaru powodzi w zlewni Górnej Wisły na tle powodzi historycznych. In: Maciejewski M, Ostojski MS, Walczykiewicz T (eds) Dorzecze Wisły: monografia powodzi maj–czerwiec 2010. IMGW-BIP, Warszawa, pp 63–71

Dubicki A (1999) Przebieg powodzi w dorzeczu górnej i środkowej Odry. In: Dubicki A, Słota H, Zieliński J (eds) Dorzecze Odry: monografia powodzi lipiec 1997. IMGW, Warszawa, pp 61–81

Dubicki A, Malinowska-Małek J (1999) Wysokość, natężenie i przestrzenny rozkład opadów atmosferycznych. In: Dubicki A, Słota H, Zieliński J (eds) Dorzecze Odry: monografia powodzi lipiec 1997. IMGW, Warszawa, pp 23–43

Dubicki A, Borowicz A, Turzańska-Chrobak B, Gierczak J, Lisowski J (1999a) Rzeczowe i finansowe straty powodziowe. In: Dubicki A, Słota H, Zieliński J (eds) Dorzecze Odry: monografia powodzi lipiec 1997. IMGW, Warszawa, pp 175–190

Dubicki A, Słota H, Zieliński J (eds) (1999b) Dorzecze Odry: monografia powodzi lipiec 1997. IMGW, Warszawa

Egler R (2003) System ochrony przeciwpowodziowej kraju, cz. 1. Gospodarka Wodna 4:143–148

Franczak P (2013) Pamięć o katastrofalnych powodziach zapisana w lokalnych legendach. Płaj 47:179–184

Franczak P (2015) Hydrologiczne i geomorfologiczne skutki ekstremalnego opadu w maju 2014 roku w zlewni Rybnego Potoku (Masyw Babiej Góry). Annales Universitatis Mariae Curie-Skłodowska. Sectio B, Geographia, Geologia, Mineralogia et Petrographia 70(2):63–81

Franczak P, Listwan K (2015) Ryzyko powodziowe w małych zlewniach górskich a sposoby zagospodarowania obszarów zalewowych zapisane w aktach planistycznych: studium przypadku Makowa Podhalańskiego i Kasinki Małej. In: Liro J, Liro M, Krąż P (eds) Współczesne problemy i kierunki badawcze w geografii, vol 3. IGiGP UJ, Kraków, pp 45–61

Franczak P, Listwan-Franczak K (2016) Powódź w zlewni Miedzianki (zlewnia Nysy Łużyckiej) w sierpniu 2010 roku. Dobra praktyka w redukcji ryzyka powodziowego w małych zlewniach górskich, w których wystąpiła powódź błyskawiczna. In: Franczak P, Krąż P, Liro J, Liro M, Listwan-Franczak K (eds) Współczesne problemy i kierunki badawcze w geografii, vol 4. IGiGP UJ, Kraków, pp 55–84

Jakóbik K (2011) Powódź w województwie małopolskim w 2010 roku. Urząd Statystyczny w Krakowie, Małopolski Ośrodek Badań Regionalnych, Kraków

Klejnowski R (2012a) Komentarz synoptyka z dnia 4 lipca 2012 r. http://weather.icm.edu.pl/komentarze/index1.php?date=2012-07-04 (10.10.2016)

Klejnowski R (2012b) Komentarz synoptyka z dnia 5 lipca 2012 r. http://weather.icm.edu.pl/komentarze/index1.php?date=2012-07-05 (10.10.2016)

Kundzewicz ZW, Matczak P (2010) Zagrożenia naturalnymi zdarzeniami ekstremalnymi. Nauka 4:77–86

Lach J (2002) Przyrodnicze i gospodarcze skutki powodzi w lipcu 2001 roku na Sądecczyźnie. Rocznik Sądecki 30:102–124

Lach J, Lewik P (2002) Powódź w lipcu 2001 na Sądecczyźnie i jej skutki. In: Górka Z, Jelonek A (eds) Geograficzne uwarunkowania rozwoju Małopolski. IGiGP UJ, Kraków, pp 199–204

Lipińska EJ (ed) (2011) Powódź 2010—przyczyny i skutki. Wojewódzki Inspektorat Ochrony Środowiska w Rzeszowie, Rzeszów. http://www.wios.rzeszow.pl/publikacje/publikacje-o-stanie-srodowiska/inne-publikacje/powodz-2010-przyczyny-i-skutki/ (10.06.2016)

Maciejewski M (ed) (2000) Model kompleksowej ochrony przed powodzią w obszarze dorzecza górnej Wisły na przykładzie województwa małopolskiego. OKI RZGW, Kraków

Maciejewski W, Ostojski MS, Tokarczyk T (eds) (2011) Monografia powodzi 2010. Dorzecze Odry. IMGW-PIB, Warszawa

Magnuszewski A, Porczek M (2015) Wskaźnik potencjału powodziowego i względna ekspozycja na niebezpieczeństwo powodziowe gmin w Polsce. Prace Stud Geogr 57:55–65

Müller U, Walther P (2014) Das Neisse-Hochwasser 2010—Analyse und Konsequenzen. In: Heimerl S, Meyer H (eds) Vorsorgener und nachsorgender Hochwasserschutz. Ausgewählte Beiträge aus der Fachzeitschrift Wasserwirtschaft. Springer-Verlag, pp 12–18

Nachlik E, Kundzewicz ZW (2016) History of floods on the Upper Vistula. In: Kundzewicz ZW, Stoffel M, Niedźwiedź T, Wyżga B (eds) Flood risk in the Upper Vistula Basin. Springer, pp 279–292

Olędzki R (2009) Komentarz synoptyka z dnia 24 czerwca 2009 r. http://weather.icm.edu.pl/komentarze/index1.php?date=2009-06-24 (10.10.2016)

Olędzki R (2012a) Komentarz synoptyka z dnia 2 lipca 2012 r. http://weather.icm.edu.pl/komentarze/index1.php?date=2012-07-02 (10.10.2016)

Olędzki R (2012b) Komentarz synoptyka z dnia 3 lipca 2012 r. http://weather.icm.edu.pl/komentarze/index1.php?date=2012-07-03 (10.10.2016)

Olszowicz A, Salamonik S, Słomska A (1999) Interpretacja synoptyczna sytuacji meteorologicznej. In: Dubicki A, Słota H, Zieliński J (eds) Dorzecze Odry: monografia powodzi lipiec 1997. IMGW, Warszawa, pp 13–22

Pociask-Karteczka J, Żychowski J (2014) Powodzie błyskawiczne (flash floods)—przyczyny i przebieg. In: Ciupa T, Sulikowski R (eds) Woda w mieście. IG UJK, Kielce, pp 213–226

*Powódź stulecia* (2010) Wiadomości Brzosteckie 6(158):8–9

*Powódź w Bogatyni. Tragiczne obrazy, żywe wspomnienia* (2012) Wydział Komunikacji Społecznej Urzędu Miasta i Gminy w Bogatyni, Bogatynia

*Powódź w gminie Czarna* (2009) OSP Czarna. http://www.ospczarna.pl/aktualnosci/powodz-w-gminie-czarna/ (8.08.2016)

*Powódź w wymiarze fizycznym, społecznym i gospodarczym* (2001) http://www.krakow.rzgw.gov.pl/index.php?option=com_content&view=article&id=1045:powod-w-wymiarze-fizycznym-spoecznym-igospodarczym&catid=103:powodzie&Itemid=360&lang=pl (5.10.2016)

*Program Odbudowa 2001* (2003) Polaka Agencja Rozwoju Przedsiębiorczości, Warszawa. https://poig.parp.gov.pl/files/74/81/88/odbudowa2001.pdf (10.10.2016)

*Promesa* (2015) UM Wojcieszów. http://www.wojcieszow.pl/portal/index.php?option=content& task=view&id=1270&Itemid=2 (10.12.2015)

*Raport po powodzi z maja i czerwca 2010* (2010) UM Krakowa, Kraków. https://www.bip.krakow. pl/plik.php?zid=76413&wer=0&new=t&mode=shw (3.07.2016)

*Raport z przebiegu powodzi w 2010 roku* (2011) Niepublikowany materiał udostępniony przez Urząd Miejski w Bieruniu

Rybak T (2011) Informacja o klęsce powodzi w 2010 roku. In: Raport o stanie środowiska w 2010 roku. WIOŚ Rzeszów, Rzeszów, pp 149–157

Salata R, Sondaj L, Wojciechowski M (2012) Ocena aktualnego stanu zabezpieczenia KWB Turów w aspekcie zdarzeń powodziowych zaistniałych w sierpniu 2010 roku. Zeszyty Naukowe Instytutu Gospodarowania Surowcami Mineralnymi Ener PAN 82:157–172

Śmiech A (2012) Sytuacja hydrologiczno-meteorologiczna w zlewni Górnej Wisły. In: Maciejewski M, Ostojski MS, Walczykiewicz T (eds) Dorzecze Wisły: monografia powodzi maj–czerwiec 2010. IMGW-BIP, Warszawa, pp 47–50

Sondaj L, Milkowski D, Wojciechowski M (2011) Zagadnienia ochrony i dostępu do zasobów węgla brunatnego w złożu Turów w aspekcie zdarzeń powodziowych zaistniałych w sierpniu 2010. Górnictwo Geoinżynieria 35(3):331–342

*Sprawozdanie—opis "Powódź w Gminie Szczucin"* (b.d.) Niepublikowany materiał udostępniony przez Urząd Miasta i Gminy Szczucin

*Sprawozdanie dotyczące prowadzonej bezpośredniej ochrony przed powodzią w dniach 7 sierpnia—1 września 2010 r.* (b.d.) Niepublikowany materiał udostępniony przez Urząd Miasta i Gminy w Bogatyni

*Sprawozdanie z działalności prezydenta miasta w czasie akcji powodziowej w okresie od 19.05.2010 r. do 24.06.2010 r.* (2010) Urząd Miasta Tarnobrzega, Tarnobrzeg

Stachý J, Fal B, Dobczyńska I, Hołdakowska J (1996) Wezbrania rzek polskich w latach 1951–1990. Materiały Badawcze IMGW Ser Hydrol Oceanol 20:79

*Stan ochrony przeciwpożarowej na terenie miasta i powiatu tarnobrzeskiego—realizacja zadań służbowych w roku 2010* (2011) Komenda Miejska Państwowej Straży Pożarnej w Tarnobrzegu, Tarnobrzeg

Tokarczyk T (2011) Przyczyny i skutki powodzi 2010 w zlewni Nysy Łużyckiej. In: Maciejewski M, Ostojski MS, Tokarczyk T (eds) Dorzecze Odry: monografia powodzi 2010. IMGW-BIP, Warszawa, pp 75–86

Tomica L, Żuławska D (2002) Wykorzystanie analiz przestrzennych do oceny skutków powodzi. http://twiki.fotogrametria.agh.edu.pl/pub/PraceMagisterskie/WebHome/PD_D_Zulawska_L_ Tomica_2002.pdf (2.08.2014)

Trybała K, Przywarska R (2004) Program ochrony środowiska dla powiatu suskiego na lata 2004–2007 wraz z perspektywą do 2011 roku. Gliwice

*Wdrażanie Dyrektywy Powodziowej w Polsce—wpływ na planowanie i zagospodarowanie przestrzenne* (2013) http://www.kzgw.gov.pl/files/file/Wiadomosci/Komisja_wspolna.pdf (21.11.2016)

Wiejaczka Ł, Bochenek W (2013) Przekształcanie dna koryta rzeki górskiej w czasie dużych wezbrań na przykładzie Ropy. Prace Geogr 132:27–38

*Wojewódzki raport z akcji przeciwpowodziowych sierpień–wrzesień 2010* (2010) Dolnośląski Urząd Wojewódzki, Wrocław

*Wspólny polsko-niemiecko-czeski raport dot. zdarzenia powodziowego w dniu 7–10 sierpnia 2010 na rzece Nysie Łużyckiej jako element wstępnej oceny ryzyka powodziowego zgodnie z art. 4 Dyrektywy Powodziowej (2007/60/EG)* (2010) http://umwelt.sachsen.de/umwelt/wasser/ download/06-12-10_LN_pol.pdf (22.08.2016)

*Wystąpienie pokontrolne (MOPS w Bieruniu) LKA-4101-02-02/2011/P/11/176* (2011) Najwyższa Izba Kontroli, Delegatura w Katowicach

*Wystąpienie pokontrolne (Urząd Miasta i Gminy Szczucin) P/13/077 LKR—4101-02-03/2013* (2013) Najwyższa Izba Kontroli, Delegatura w Krakowie

*Wystąpienie pokontrolne (Urząd Miasta Tarnobrzega) P/15/097 LRZ.410.004.05.2015* (2015) Najwyższa Izba Kontroli, Delegatura w Rzeszowie

*Wystąpienie pokontrolne (Urząd Miejski w Brzostku) P/13/077 LRZ-4101-04-01/2013* (2013) Najwyższa Izba Kontroli, Delegatura w Rzeszowie

*Wystąpienie pokontrolne (Urząd Miejski w Pilźnie) P/15/097 LRZ.410.004.13.2015* (2015) Najwyższa Izba Kontroli, Delegatura w Rzeszowie

Zawiślak T, Adamczyk Z, Bąkowski R (2012) Synoptyczne uwarunkowania powodzi. In: Maciejewski M, Ostojski MS, Walczykiewicz T (eds) Dorzecze Wisły: monografia powodzi maj–czerwiec 2010. IMGW-BIP, Warszawa, pp 13–26

# Chapter 3
# Changeability of the Spatial Image of Flood Over Time

If you can see a river out of your window then, sooner or later, you should expect to suffer the consequences of a flood. Not every flood will pose a direct threat to your house, but it will affect your life indirectly. This kind of an opinion is often heard from people living in areas that have experienced flooding. The research was based on methods useful in collecting data as close to the intuitive perception of a flood as possible, drawing on the long-established concept of the image, which, according to Boulding's theory (1956), people build throughout their lives based on information obtained from the surrounding social and physical environment. Research has revealed that people's responses to natural hazards are a result of their choosing between different images of natural disasters, but the choice itself is only a partly rational process, and it is always based on incomplete and presumably inaccurate information (Kates 1962; Johnston 1979). It should be noted that the incompleteness of information follows both from objective absence of such information and the nature of human cognitive processes.

Cognitive mapping is a process of internal psychological transformation during which each person creates representations of the relative locations and attributes of their everyday environment (Downs and Stea 1973). Cognitive maps are created by drawing a sketch on a base map with places meeting certain criteria, for example, that are at risk of flooding (Ruin et al. 2007). They represent a cartographic record of individual perception of risk (O'Neill et al. 2015). Unfortunately, these images are not perfect representations of human spatial perception because there may be discordance between what each person knows about a given area and his or her ability to render this knowledge on a map (Didelon et al. 2011). However, because of the advantages of this method, cognitive mapping has been utilised in various studies on the perception of different components of the human environment, e.g. natural hazards such as volcanoes (Gaillard 2008; Gaillard et al. 2001; Leone and Lesales 2009) and floods (Brilly and Polic 2005; Pagneux et al. 2011; Ruin et al. 2007), and in studies focusing on the perception of anthropogenic hazards, e.g. crime (Curtis 2012; Curtis et al. 2014; Matei et al. 2001) and pollution (Brody et al. 2004; Stone 2001). In this study, we understand a cognitive map as the cartographic representation of

© The Author(s), under exclusive licence to Springer Nature Switzerland AG 2019    43
J. Działek et al., *Understanding Flood Preparedness*, SpringerBriefs in Geography,
https://doi.org/10.1007/978-3-030-04594-4_3

the area considered as flooded and drawn by inhabitants of localities that suffered from floods several years before our field study.

There is a consensus as to the impact of one's experience on the perception of the flood risk (Felgentreff 2003; Grothmann and Reusswig 2006; Plapp and Werner 2006; Siegrist and Gutscher 2006; Miceli et al. 2008; Heitz et al. 2009; Terpstra 2009). Research has further demonstrated that the perception and awareness of the risk reach high levels immediately after the flood, but they disappear soon thereafter (Felgentreff 2003). The fact that humans forget is an important factor in the cognitive process since—in addition to triggering other processes, such as the cognitive dissonance—it can lead to significant changes in the knowledge about a phenomenon (Biernacki et al. 2008). The subjective memory of a community member influences the 'social memory of a flood' (Komac 2009). As it changes over time, the image of a flood modifies a person's readiness to assimilate information about the phenomenon, and as a consequence, time strongly diversifies the willingness of community members to take protective action against the consequences of potential floods (Kellens et al. 2011; Adger et al. 2009).

## 3.1    The Case of Bieruń

The team of researchers visited the inhabitants of Bieruń 4 years after the flood. After such a timespan, most respondents were able to give the exact date of the flood (18 May 2010). Likewise, nearly all respondents remembered the developments which led to the flood, namely, strong rainfall and the breach of the flood bank (the residents were not unanimous as to whether this resulted from its structure having been weakened and the pressure of water, or from intentional, covert human action). For most of those surveyed, the flood lasted until the water subsided and they were able to start putting their farmyards into order and restore their daily routines. It should be noted that in the case of Bieruń, water remained behind the levees for several weeks, and for some persons, the time was so difficult that they now perceive the flood as lasting until 'the end of the summer holidays' (end of August), when most of the damage caused by the floods and the 'mementos' it had left behind were removed from their closest surroundings.

No one expected the water to come from that direction.

You have no idea what it's like when you come back from work and can only see your rooftop.

The impossible became possible.

Most of the respondents best remember the fight against the inflowing water, the scarce time left for rescuing their belongings and animals, and the stacking of sandbags, which did not prove to be as effective as it should have been. The floodwaters inundated a vast area and were even several metres deep, as a consequence of which transport by boat was the only feasible option and a much needed one. In addition, the police were required to patrol the flood area and guard the empty houses. The residents associate the earliest hours of the flood with dedicated actions on the part of the emergency teams—the firefighters and the police. The hours and days which followed are very often remembered for the trauma and stress they caused, as well as the smell coming from the stagnant waters.

Some of the flood victims tend to engage in talking about the flood and remembering it at times of heavy rain.

The topic is brought up again when someone is ill or when it rains.

It all comes back during rain, it is terrible.

However, persons seriously affected by the flood, who represented a large proportion of the interviewees, were clearly negating the sense of remembering, analysing and collecting information.

We do not talk, we do not want to discuss it, so that we can forget.

I don't remember anymore, I have forgotten. You try to push it out of your mind.

We avoid such conversations.

As regards Bieruń, there is far-reaching concurrence (Fig. 3.1) between how the inhabitants perceived the flood and the actual flood extent. The topographical conditions, the type of building development within the areas flooded and the course of the flood—the fact that residents were able to watch the stagnant water for a longer period—made the image of the flood area stay in their memory.

Differences in terms of indications are observable between the eastern and western areas of Bieruń. Even though the central part of the overflow area was marked by all the respondents, the inhabitants of the eastern end of Bieruń tended to mark that part of it more eagerly, while those from the western part marked the opposite end.

When comparing the reach of the flood marked by women and men, it can be seen that neither its location nor its size is gender-specific. The smallest area indicated by a respondent implies how idiosyncratic knowledge about the extent of a flood can be. In 2014, an area of 1.13 km$^2$ was identified as flooded by one of the respondents, which represented 31% of the actually affected area (Fig. 3.2), while in 2016, it was 0.46 km$^2$ (13%). The average surface area considered by the respondents to have been struck by the flood also tends to shrink over time: it was 2.7 km$^2$ in 2014 and 2.2 km$^2$ in 2016.

The figures relating to the memory of individual persons provide interesting information. As mentioned above, the differences between the reach of the flood as such are not significant. Therefore, the surface areas were compared. First, there were differences between individual respondents, reaching 66% in 2014, and as much as 86% in 2016. In addition, certain tendencies are observable as regards changes in the flood image over time. First of all, smaller areas were marked during the second round of the interviews. Furthermore, the larger the area indicated during the first interview, the smaller the reduction in size after the 2-year period. However, there are two cases deviating from this rule, where despite the quite high accuracy in 2014, a drop in the area of more than 70% was recorded. The above data demonstrates that, over time, memory of events—even those having a huge emotional load—tends to become poorer in a natural way. Obviously, there are exceptions from this rule, as was indicated above.

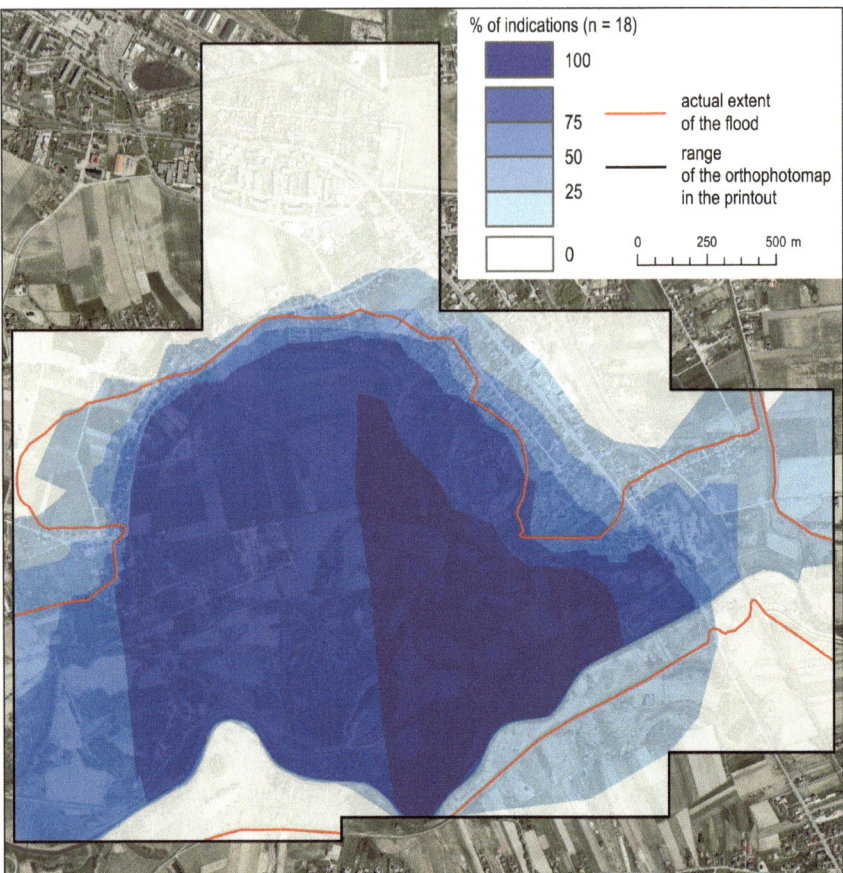

**Fig. 3.1** Flood areas in Bieruń as indicated by the respondents (2014). *Note* The actual boundaries of the flood in all the maps presented in this chapter were delimited by P. Franczak during the field surveys. *Source* Own study

Equally interesting information is provided by a comparison of the distance between one's home and the flood area delimited by the respondents (the home representing its centre of gravity—the centroid) with the surface area of the affected land delimited by the respondents. The determination of the centre of gravity is considered to be a means of transforming inaccurate, incalculable qualitative information associated with indescribable human feelings into quantitative data which can be used in subsequent statistical analyses (Hurtado 2010). By assumption, as the abovementioned distance increases, which also means better familiarity with the area affected by the flood, the surface area in question should grow bigger. Furthermore, the tendency should become even clearer over time. It can be concluded that the case of Bieruń is a model example corroborating the above theory (Fig. 3.3). Naturally, statistics based on such a small study sample have clear imperfections, but the results

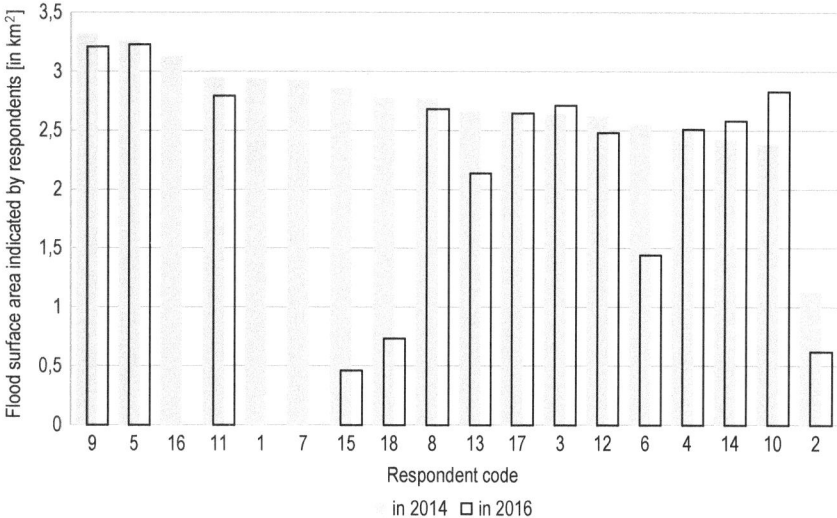

**Fig. 3.2** Flood area indicated by respondents in Bieruń in 2014 and 2016. *Source* Own study

obtained do not, by any means, disprove the above regularity. Importantly, for each of the characteristics, eliminating the extreme results does not significantly change the established trend.

The last category of numerical data analysed in detail is the correlation between the distance from the respondent's house to their farthest point of activity and the surface area of the flood. By analogy to the above trend, it was assumed that the more active an individual was, the larger (or more accurate) area they would indicate, with the tendency becoming more evident over time. For Bieruń, the significance of the passing of time is confirmed (Fig. 3.4). However, the results of the first interview show a reverse tendency to that assumed. Perhaps, the accuracy in question should not be attributed to the surface area of the flood indicated. However, we decided to verify whether disregarding the extreme result would materially alter the overall outcome. This was indeed the case. One of the very active persons marked a relatively small flood area during the first interview (the smallest of all). In the other cases, the hypothetical trend is not very substantial, but it does nevertheless occur and, as above, does not disprove the assumption made above.

## 3.2 The Case of Czarna

When surveyed 5 years after the flood, most inhabitants were unable to give its exact date (29 June 2009). Some believed it had been the beginning of summer holidays, some thought it was late June or early July 2009 and some even declared 2010. The

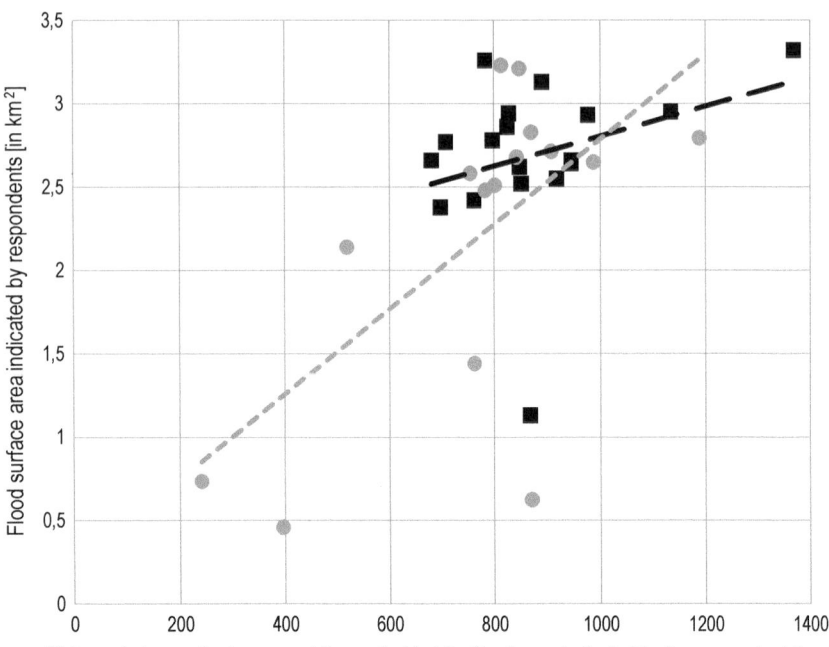

**Fig. 3.3** Relationship between the distance from the house to the centroid of the flood area and the surface area of the flood for individual respondents. *Source* Own study

flood lasted 2–3 days, following which water returned to the river channel, ceasing to pose a threat and nuisance to the inhabitants. On account of the destruction and need to renovate their farm buildings, two persons extended the duration of the flood to a month.

For most of the inhabitants, water had arrived from the forest after heavy rainfall, but they could not tell how and when this happened. Following this, it began to rise abruptly. Several persons living near small streams on both sides of the river also saw them swell, and eventually flood the nearby areas. A very characteristic landmark for the local population was the bridge that connects the village on the two banks of the river. The river level rose so high, inter alia on account of the small clearance under the bridge, that it flooded the areas of the nearby kindergarten, swimming pool, school and playground. Soon thereafter, news about imminent flood wave started to spread, which was remembered well by the respondents.

The people of Czarna do not declare having any major problems returning to the memories of the event. The scale of the damage, loss, hindrance and duration was much smaller than in Bieruń, which was described above. Heavy, long-term rainfall in the area is the most frequently mentioned trigger of memories of the flood.

I'm scared whenever it starts raining heavily.

Sometimes I walk there and tell the river that I like her, but I am afraid of her anyway.

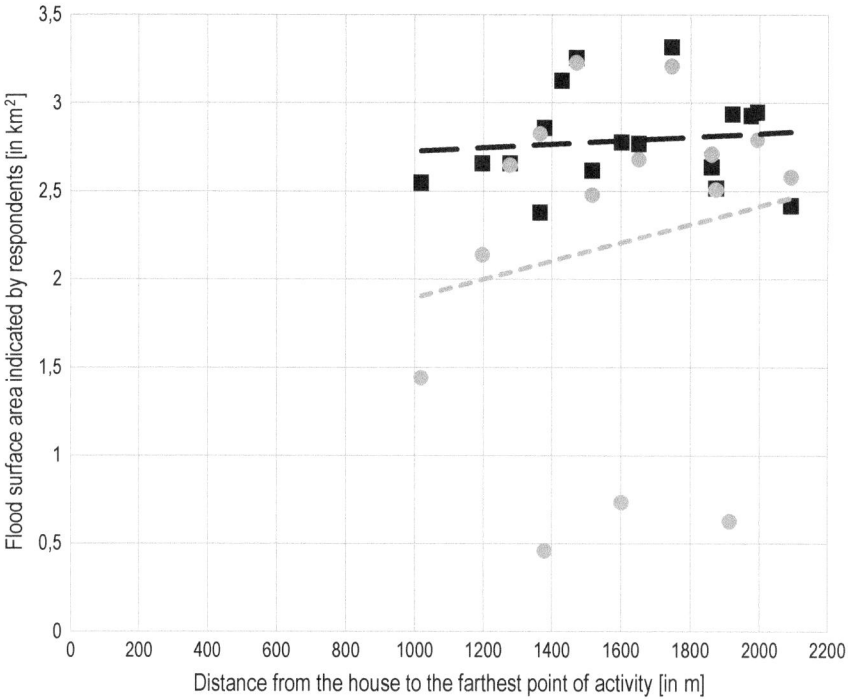

**Fig. 3.4** Relationship between the distance from the house to the farthest point of activity and the surface area for individual respondent. *Source* Own study

Czarna demonstrates a slightly poorer convergence between the image of the flood and its actual extent, especially in its western part (Fig. 3.5). Residents quite accurately indicate the areas in the vicinity of the bridge, but do not have the memory of the areas to the west. This is not surprising because they are, for the most part, covered by forests or farmland, so accurate delimitation of the flood boundaries is much more difficult there, and much less interesting given the value of those terrains.

The differences between the indications of the residents living in the western and eastern parts of Czarna, respectively, were most pronounced for the western forest and agricultural land mentioned above. Those who live closer declare a more accurate range of the floodwaters.

When comparing the extent of the floods marked, similar conclusions as for Bieruń can be drawn. The differences between men and women are negligible, while—notwithstanding some similarities in the perception of landmarks—the overall memory of the flood is highly individualised. More pronounced differences can be observed in the size of the areas indicated, especially the smallest ones. In 2014, the smallest area was 0.10 km², representing 29% of the area actually affected by the flood (Fig. 3.6), while in 2016—0.01 km² (3%). With time, the average size of

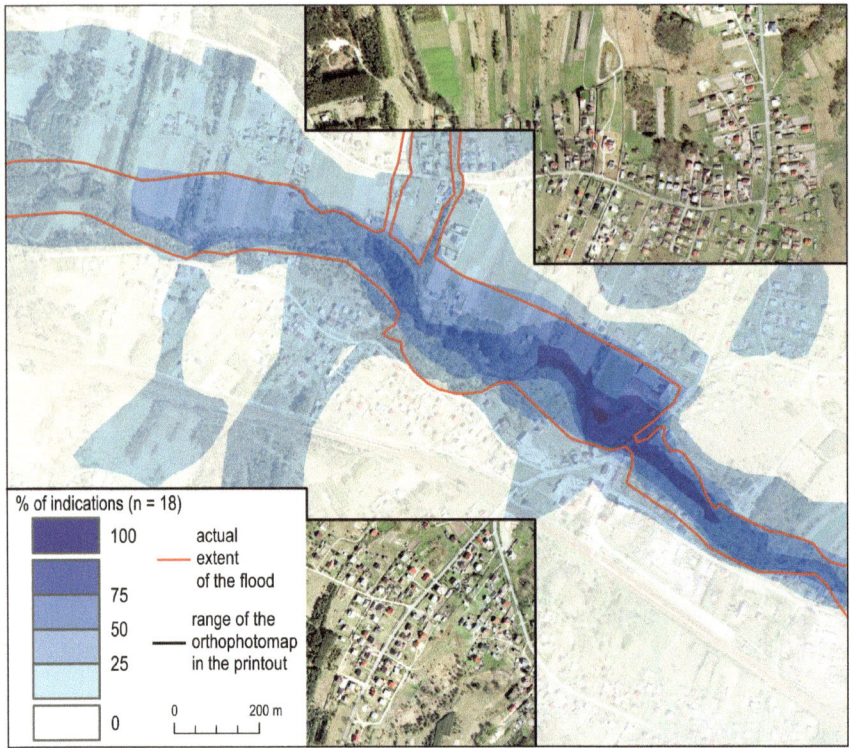

**Fig. 3.5**  Flood areas in Czarna as indicated by the respondents (2014). *Source* Own study

the area considered to have been affected by the flood decreased significantly: from 0.22 km$^2$ in 2014 to 0.13 km$^2$ in 2016 (by 41%).

When studying the memory of the population of Czarna, interesting conclusions can be drawn after comparing the size of the flood areas indicated by individual respondents over time, as well as the areas as marked in general. There is a distinguishable tendency on the part of those surveyed to indicate smaller areas. Some people identified a slightly larger area (from 4 to 41%). The flood shrank in the residents' minds from 4 to as many as 92%, which can be interpreted as near complete erasure of the flood from their memory.

Major differences are also noticeable between the respondents' indications within the same round of interviews. The gaps between the maximum and minimum sizes were 82 and 97% in 2014 and 2016, respectively. This confirms the conclusion based on the Bieruń case study—the image of a flood in human memory highly deteriorates with time. In addition, the deterioration is the stronger the less seriously the disaster affects a given area.

The comparison of the distance between the house and the centroid of the flood area delimited by the respondents produced utterly different outcomes for Czarna than for the 'model' case of Bieruń. While the indications from the first interview

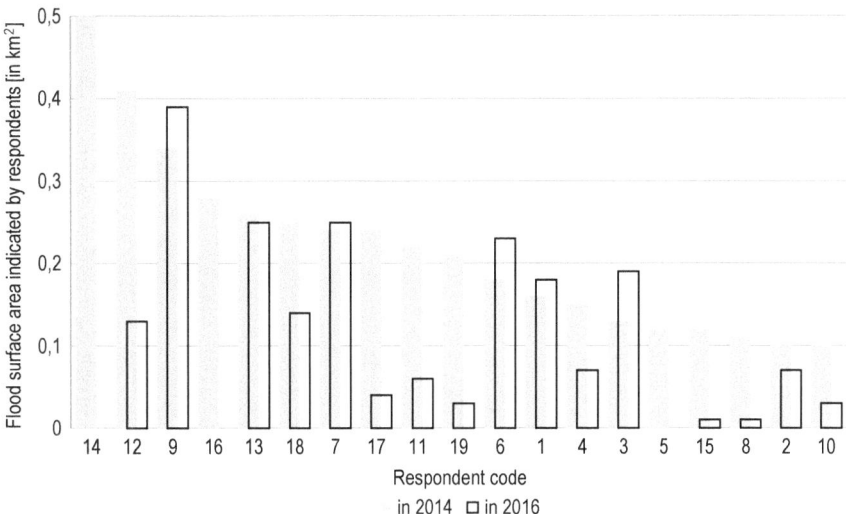

**Fig. 3.6** Flood area indicated by respondents in Czarna in 2014 and 2016. *Source* Own study

confirmed the assumptions, those from 2016 contradicted them (Fig. 3.6). As a consequence, consideration should be given to what causes the anomaly: why do the people who indicated areas highly distanced from their homes tended to reduce them so considerably? This can be attributed to two factors. First, the shape of the area in question, as well as its relief, and shape of the river valley resemble a circle to a much lesser extent than in Bieruń. This explains why a smaller area in general is marked. The difference over time can also be connected with the trend described above—the residents of Czarna were not affected as much as in Bieruń, and thus their image of the flood is different, less intense and less accurate. Even though they are able to position the area of the disaster in space, they narrow it down over time. The place is familiar to them (hence the far distance from their houses), but they do not remember the exact reach of the water (hence the reverse trend).

A similar effect, which supports concluding that Czarna residents push the flood out of their awareness, is implied by the relationship between the distance from home to the farthest point of the respondent's activity, on the one hand, and the size of the flood area, on the other (Fig. 3.7). The outcome should disprove the above trend. However, this is not actually the case. The flood in Czarna was less intense than in Bieruń, which had a strong enfeebling effect on the memory of the disaster and increased the divergence between the perception by the respondents and the facts (Fig. 3.8).

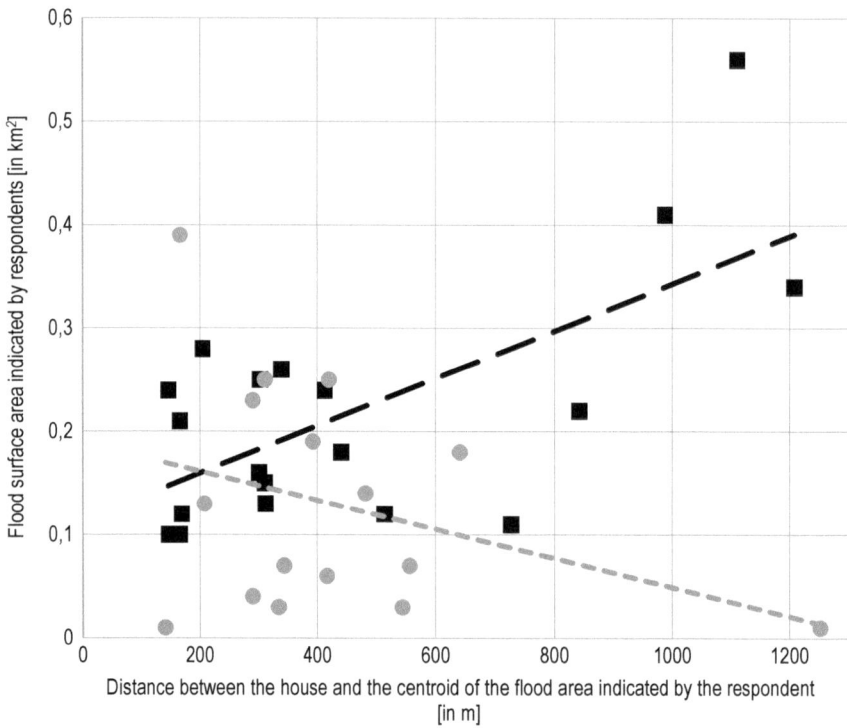

**Fig. 3.7**  Relationship between the distance from the house to the centroid of the flood area and the surface area of the flood for individual respondents. *Source* Own study

## 3.3  The Case of Wojcieszów

During the interviews 2 years after the flood, residents have much difficulty giving the exact start date of the flood, with most of them remembering it was 'the beginning of the summer holidays' in 2012. When water returned to the river channel and residents were able to move around the town again, the flood, as remembered by the inhabitants, was over—this was up to one week.

The flood in Wojcieszów was by far the most violent one of all the floods described in this chapter. As the inhabitants remember it, there were recurrent downpours, water began to come down from the surrounding mountains, following which its level began to rise in the river, which was highly unexpected and had major implications for the town since it is located in a river valley, with the road ensuring a transport link for the longitudinally stretched Wojcieszów corresponding to the axis of the river. For this reason, the residents would repeatedly refer to the showers, the mountains, the river and the road in their accounts, and clearly associated them with the flood. They slightly less frequently recalled the small walls which are part of the river

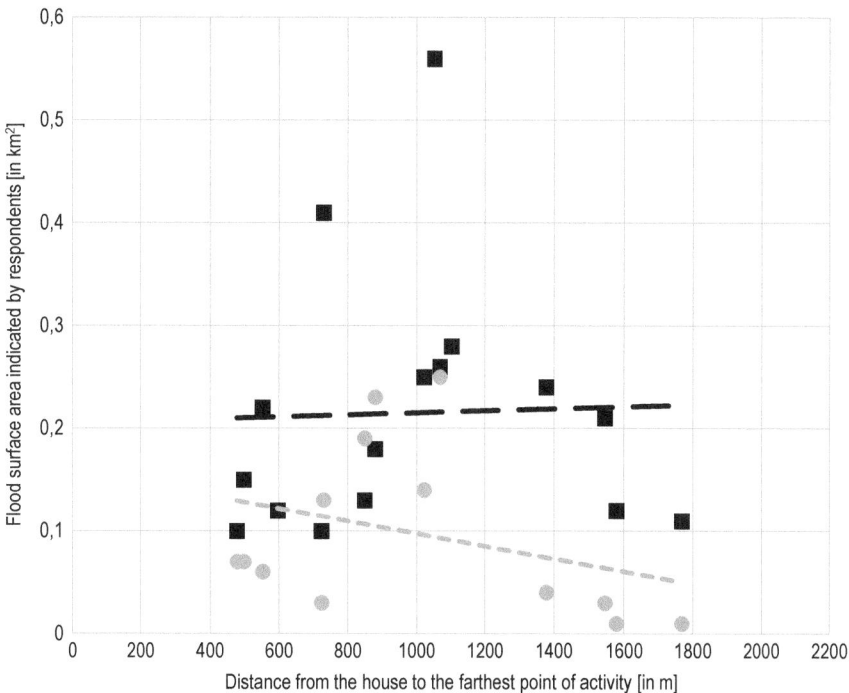

**Fig. 3.8** Relationship between the distance from the house to the farthest point of activity and the surface area for individual respondents. *Source* Own study

embankments, the damaged bridge and the most frequent damage suffered by the households, i.e. flooded basements.

As in the other towns, recurrent heavy rainfall brings back people's memories, and stimulates conversation about the flood they experienced. Most probably, the abrupt nature of the flood and the associated sense of its unpredictability discourage people from talking about the flood. The prevailing opinion is that it will not reoccur on such a scale.

No need to come back to it over and over again, it just came and went, nothing could stop it.

In our area, people only had their basements flooded, the road was ruined, but I do not wish it on anyone, anyway.

Of the three areas surveyed, the smallest correspondence between the area indicated by the respondents and the actual extent of the flood was observed for Wojcieszów. Naturally enough, most of those surveyed indicated the river and the road, but the knowledge about the other landmarks is strongly correlated with their place of residence, and was the fact which determined the degree of familiarity with the scale and extent of the flood. The significance of the road as the axis of the building development area is confirmed not only by the interviews but also by an analysis of the town's morphology, as well as the cartographic and statistical material collected.

Clearly, the area indicated by those surveyed corresponds to the actual extent of the floodwater (Fig. 3.9), but for Wojcieszów using surface statistics is slightly less warranted than in the other cases.

Wojcieszów stretches longitudinally, as a result of which the indications by the inhabitants differed depending on where they lived: in north or in the south. The information obtained from the residents demonstrated the strongest relationship with the location of the observation point among all the cases analysed here. It turns out that none of the nine respondents living in the northern part of the town indicated the area of the school complex in the southern part. By contrast, the areas along the river channel in the northern part were marked quite accurately. The residents of the southern part presented a slightly different tendency. Naturally, they delimited the flood area near their own place of residence more accurately, and tended to overstate the areas affected by floodwater for more distant locations.

No gender-based differences in the flood areas marked by the respondents are noticeable for Wojcieszów, either. On the other hand, the indications of the flood area in the town demonstrate the farthest-reaching individualisation. In 2014, the smallest area identified covered 0.03 km$^2$, which represented 9% of the area actually affected by the flood, while in 2016 it was 0.01 km$^2$ (3%). The average size of the affected area marked is nearly identical in both surveys. However, the figures differ significantly from one person to another, with the average outcome highly influenced by one answer. Without it, the average size of the area in question decreased in the memory of the local population by 44% over the 2-year period (Fig. 3.10).

It is clear from a comparison of the flood areas indicated by the different respondents that, over time, the area they remember tends to shrink significantly, which means that the flood blurs in their memory. This is a true conclusion, even though the average figures presented above disprove it. This is attributable to the answer of respondent 2, which highly diverges from the other ones (falling out of the graph's scale). As for Czarna, major differences are noticeable between the respondents' indications within the same round of interviews. The differences between the maximum and minimum surface area were 94 and 99% in 2014 and 2016, respectively. The above data also confirms the finding that the image of flood people has in their memory fades over time.

The relationship between the distance from the house to the flood and the surface area also proves to be correct for Wojcieszów. Because the town is stretched longitudinally, the distance from the place of residence to the flood area acquires particular significance. Thus, the greater the distance from the house to the centre of the flood, the higher the area marked by the respondent. In other words, the farther the range of observation by a given person, the greater the flood they see. This dependence tends to gain significance over time (Fig. 3.11).

A completely new light is shed when the distance from one's house to the farthest point of the respondent's activity is compared to the area of the flood. The trend lines (Fig. 3.12) have negative orientation, which means that the mechanisms observed for the other towns did not work. As it turns out, more active persons indicated smaller areas. This seems to disprove the effect of activity on improved memory of a flood. However, it must be remembered that the data represents the habitual activity of the

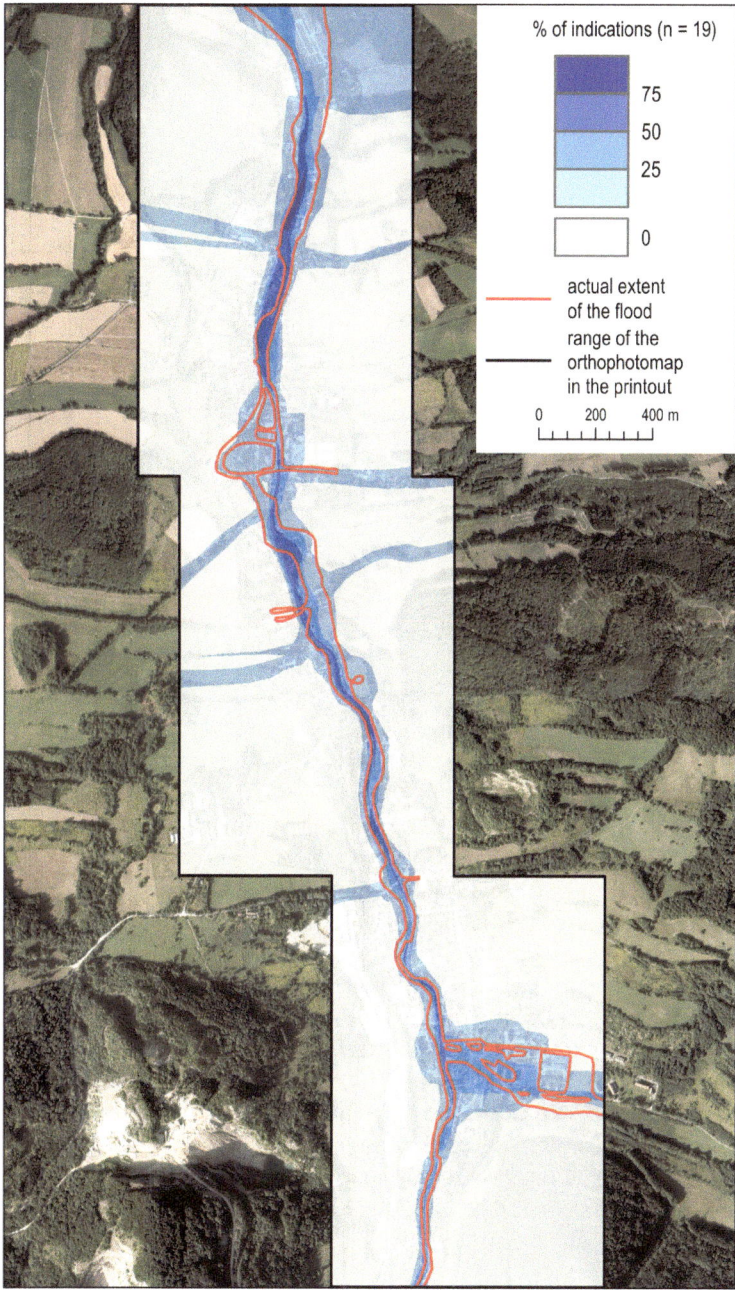

**Fig. 3.9** Flood areas in Wojcieszów as indicated by the respondents (2014). *Source* Own study

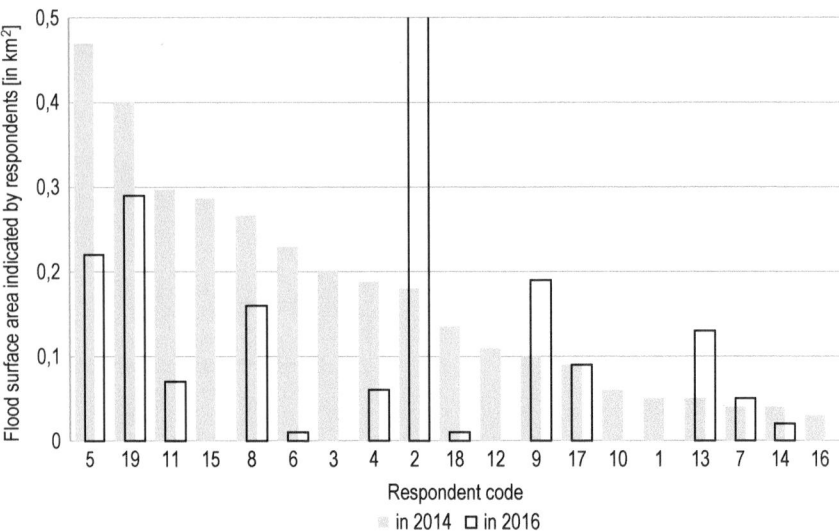

**Fig. 3.10** Flood area indicated by respondents in Wojcieszów in 2014 and 2016. *Source* Own study

inhabitants, rather than their activity during the flood. In Wojcieszów, moving about the town during the disaster was impossible due to the flooding of the road. As a consequence, the most active persons were most affected by constrained mobility, and thus gained and preserved in their memory less accurate knowledge about the phenomenon.

The representations of the environment, in this case of areas covered by floodwaters, studied within two time intervals since the occurrence of the flood, are essential in understanding human behaviour and the interactions of individuals and groups with their surroundings (Curtis et al. 2014) Significant discrepancies between the real extent of the flood zone and its perception by flood victims only 2 years after its occurrence (in all the cases studied) are an important message for individuals and organisations involved in risk education activities (McEwen and Jones 2012). Since, to the author's knowledge, there were no similar studies undertaken, no comparison could be drawn; however, we believe that the use of the cognitive mapping method could complement traditional flood risk management, also as a part of social vulnerability assessment (Reichel and Fromming 2014). Therefore, understanding how people perceive the flood, the risk of its occurrence and its possible extent based on their prior experience could be integrated into risk management policies (Correia et al. 1998; Motoyoshi 2006).

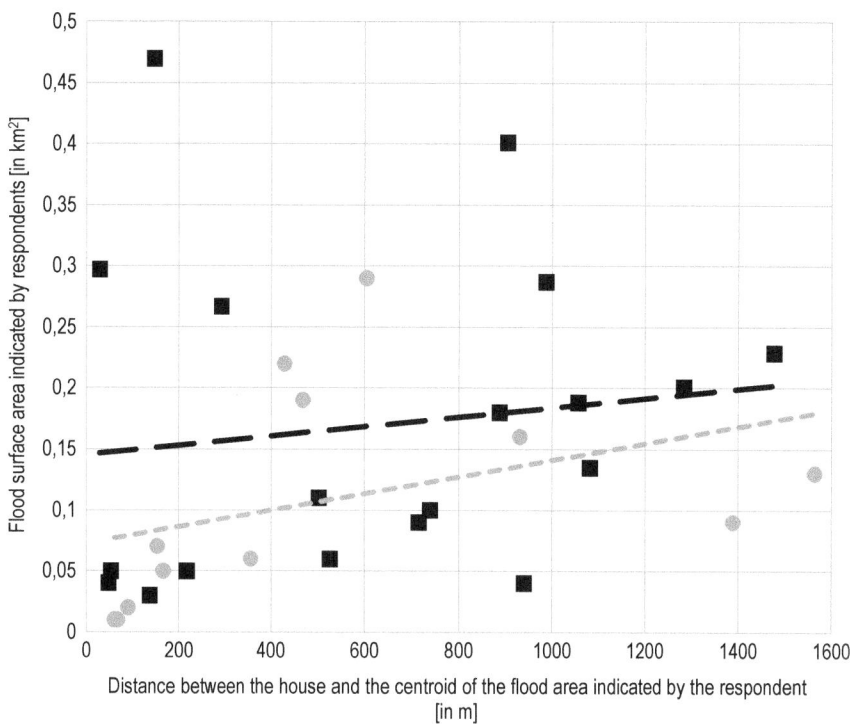

**Fig. 3.11** Relationship between the distance from the house to the centroid of the flood area and the surface area of the flood for individual respondents. *Source* Own study

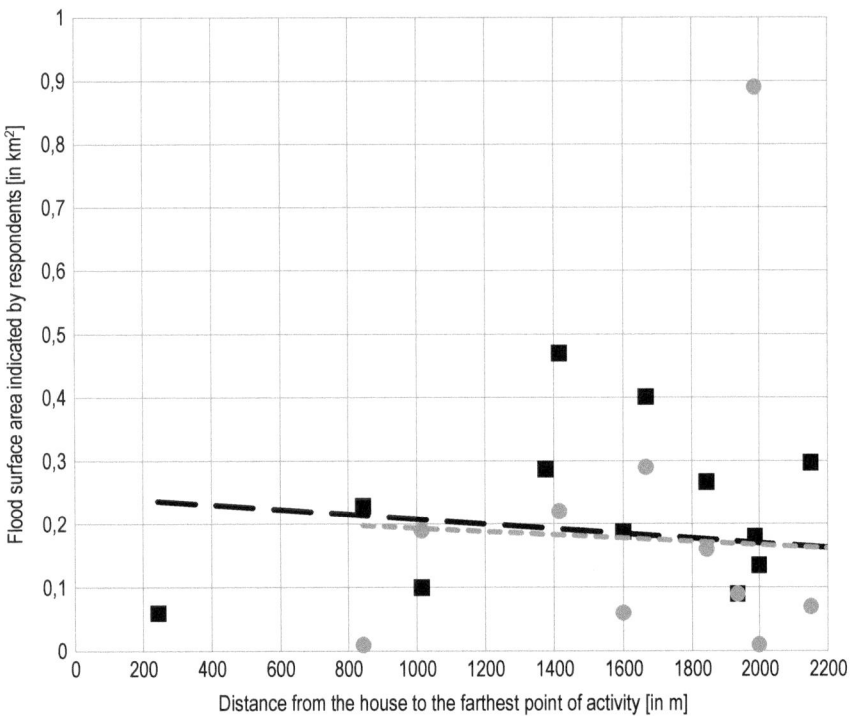

**Fig. 3.12** Relationship between the distance from the house to the farthest point of activity and the surface area for individual respondents. *Source* Own study

# References

Adger WN, Dessai S, Goulden M, Hulme M, Lorenzoni I, Nelson D, Wreford A (2009) Are there social limits to adaptation to climate change? Clim Change 93(3–4):335–354

Biernacki W, Bokwa A, Domański B, Działek J, Janas K, Padło T (2008) Mass media as a source of information about extreme natural phenomena in Southern Poland. In: Carvalho A (ed) Communicating climate change: discourses, mediations and perceptions. Centro de Estudos de Comunicação e Sociedade, Universidade do Minho, Braga

Boulding KE (1956) The image: knowledge in life and society. University of Michigan Press, Ann Arbor

Brilly M, Polic M (2005) Public perception of flood risks, flood forecasting and mitigation. Nat Hazards Earth Syst Sci 5(3):345–355

Brody SD, Peck BM, Highfield WE (2004) Examining localized patterns of air quality perception in Texas: a spatial and statistical analysis. Risk Anal 24(6):1561–1574

Correia FN, Fordham M, Saraiva MDG, Bernardo F (1998) Flood hazard assessment and management: interface with the public. Water Resour Manage 12(3):202–227

Curtis JW (2012) Integrating sketch maps with GIS to explore fear of crime in the urban environment: a review of the past and prospects for the future. Cartogr Geogr Inf Sci 39(4):175–186

Curtis JW, Shiau E, Lowery B, Sloane D, Hennigan K, Curtis A (2014) The prospects and problems of integrating sketch maps with geographic information systems to understand environmental perception: a case study of mapping youth fear in Los Angeles gang neighborhoods. Environ Plan B Plan Des 41(2):251–271

Didelon C, Ruffray SD, Boquet M, Lambert N (2011) A world of interstices: a fuzzy logic approach to the analysis of interpretative maps. Cartogr J 48(2):100–107

Downs R, Stea D (1973) Image and environment. Aldine Publishing Company, Chicago

Felgentreff C (2003) Post-disaster situations as "window of opportunity"? Post-flood perceptions and changes in the German Odra River region after the 1997 flood. Erde 134:163–180

Gaillard J-C (2008) Alternative paradigms of volcanic risk perception: the case of Mt. Pinatubo in the Philippines. J Volcanol Geotherm Res 172(3):315–328

Gaillard J-C, D'Ercole R, Leone F (2001) Cartography of population vulnerability to volcanic hazards and lahars of Mount Pinatubo (Philippines): a case study in Pasig-Potrero River basin (province of Pampanga). Géomorphol Relief Process Environ 7:209–221

Grothmann T, Reusswig F (2006) People at risk of flooding: why some residents take precautionary action while others do not. Nat Hazards 38:101–120

Heitz C, Spaeter S, Auzet AV, Glatron S (2009) Local stakeholders' perception of muddy flood risk and implications for management approaches: a case study in Alsace (France). Land Use Policy 26:443–451

Hurtado SM (2010) Modeling of operative risk using fuzzy expert systems. In: Glykas M (ed) Fuzzy cognitive maps. Studies in fuzziness and soft computing, vol 247, pp 135–159

Johnston RJ (1979) Geography and geographers: Anglo-American human geography science 1945. Arnold, London

Kates RW (1962) Hazard and choice perception in flood plain management. Research paper no. 78. Department of Geography, University of Chicago

Kellens W, Zaalberg R, Neutens T, Vanneuville W, De Maeyer P (2011) An analysis of the public perception of flood risk on the Belgian Coast. Risk Anal 31(7):1055–1068

Komac B (2009) Social memory and geographical memory of natural disasters. Acta Geogr Slov 49(1):199–226

Leone F, Lesales T (2009) The interest of cartography for a better perception and management of volcanic risk: from scientific to social representations: the case of Mt. Pelée volcano, Martinique (Lesser Antilles). J Volcanol Geotherm Res 186(3):186–194

Matei S, Ball-Rokeach SJ, Qiu JL (2001) Fear and misperception of Los Angeles urban space: a spatial-statistical study of communication-shaped mental maps. Commun Res 28:429–463

McEwen L, Jones O (2012) Building local/lay flood knowledges into community flood resilience planning after the July 2007 floods, Gloucestershire, UK. Hydrol Res 43:675–688

Miceli R, Sotgiu I, Settanni M (2008) Disaster preparedness and perception of flood risk: a study in an alpine valley in Italy. J Environ Psychol 28:164–173

Motoyoshi T (2006) Public perception of flood risk and community-based disaster preparedness. In: Ikeda S, Fukuzono T, Sato T (eds) A better integrated management of disaster risks: toward resilient society to emerging disaster risks in mega-cities. Terrapub and Nied, Tokyo, pp 121–134

O'Neill E, Brennan M, Brereton F, Shahumyan H (2015) Exploring a spatial statistical approach to quantify flood risk perception using cognitive maps. Nat Hazards 76(3):1573–1601

Pagneux E, Gísladóttir G, Jónsdóttir S (2011) Public perception of flood hazard and flood risk in Iceland: a case study in a watershed prone to ice-jam floods. Nat Hazards 58(1):269–287

Plapp T, Werner U (2006) Understanding risk perception from natural hazards: examples from Germany. Risk 21:101–108

Reichel C, Fromming UU (2014) Participatory mapping of local disaster risk reduction knowledge: an example from Switzerland. Int J Disaster Risk Sci 5:41–54

Ruin I, Gaillard J-C, Lutoff C (2007) How to get there? Assessing motorists' flash flood risk perception on daily itineraries. Environ Hazards 7(3):235–244

Siegrist M, Gutscher H (2006) Flooding risks: a comparison of lay people's perceptions and expert's assessments in Switzerland. Risk Anal 26(4):971–979

Stone JV (2001) Risk perception mapping and the Fermi II nuclear power plant: toward an ethnography of social access to public participation in Great Lakes environmental management. Environ Sci Policy 4(4–5):205–217

Terpstra T (2009) Flood preparedness: thoughts, feelings and intentions of the Dutch public. University of Twente, The Netherlands

# Chapter 4
# Social Vulnerability as a Factor in Flood Preparedness

## 4.1 The Concept of Social Vulnerability

The degree of flood preparedness comes not just from the perception of the flood risk among inhabitants of areas at risk but is also influenced by social and economic determinants referred to as social vulnerability. Indeed, the mere awareness of a potential natural disaster does not always mean that protective steps are taken due to various limitations, e.g. insufficient knowledge on how to prepare, lack of financial resources or health problems.

The increasing interest in the concept of social vulnerability follows from changes in the approach to disaster risk reduction. In the context of the flood hazard, this means transition from the traditional approach of flood protection to flood risk management (UN/ISDR 2004, 2012; Thomas et al. 2013; Raška 2015; Vávra et al. 2015; Fox-Rogers et al. 2016). The former approach focuses on the physical course of a catastrophic event, and its main purpose is to bring the forces of nature under control, to separate humans and their activities from the natural changeability of the environment by means of infrastructural solutions. The former approach is also linked with the philosophy of centralised top-down management, with a strong role for the authorities and emergency management services. This approach is opposed by an emerging paradigm based on the concepts of public participation, the use of local knowledge and skills, and shared responsibility of local populations for managing their local environment, with more respect for the natural world. It attaches greater importance to the prevailing local social, economic, cultural and political conditions (Fordham et al. 2013).

Enhancing participation and joint responsibility among vulnerable populations entails the need to establishing to what extent they are capable of handling a natural disaster. The concept of social vulnerability used in this research highlights that not all people are equally capable of adequately coping with the elements. It is derived from direct observations which indicate that the consequences of the same event may vary between individuals, groups and communities in terms of number of deaths, health

J. Działek et al., *Understanding Flood Preparedness*, SpringerBriefs in Geography,
https://doi.org/10.1007/978-3-030-04594-4_4

consequences, losses or the degree to which the foundations of a given community are shaken (Wisner 2004; Fekete et al. 2014). Thus, the concept recognises that social inequalities are reflected in the capacity of individuals, groups and communities to cope with the power of nature and the consequences it can produce (Kuhlicke et al. 2011; cf. Cutter 2013; Thomas et al. 2013). It refers to a range of societal, cultural and economic features that are designated in this study as social vulnerability drivers, which contribute to social disparities thus influencing how individuals, social groups and communities are able 'to anticipate, cope with, resist and recover from the impact of a natural hazard' (Blaikie et al. 1994: 9).

The theorists of social vulnerability to natural disasters underline the multidimensional nature of the concept (Birkmann 2006a; Tate 2012), as a construct, which is not directly observable, but which can be described by means of a number of proxy characteristics. The resulting difficulty in defining social vulnerability, its linkages with other concepts and finally, the methods for its quantification are subject to a heated scholarly debate. Researchers have already created a number of overview studies dealing with the conceptualisation (e.g. Thywissen 2006; Tapsell et al. 2010; Birkmann et al. 2013; Thomas et al. 2013) and operationalisation of social vulnerability (e.g. Birkmann 2006b; Gall 2007; Cutter et al. 2009; Fekete 2012; Tate 2012; Yoon 2012; Rufat et al. 2015). They address, inter alia, the problem of the determinants of social vulnerability being mistaken for social vulnerability as such. Thus, the concept of social vulnerability as a certain overarching concept referring to the deficits of individuals, groups or communities must be separated from the drivers of the level of social vulnerability and its potential impacts. As a consequence, this study uses the definition of social vulnerability to natural disasters according to which it is a set of diagnosable deficits specific to a given individual, group or community resulting from socio-economic disparities that may affect their capacity to cope with a natural disaster before, during and after it occurs, i.e. at the time when they prepare for the hazard, respond to it and recover from its consequences. Emphasis will be placed on the impact of social vulnerability level on the level of flood preparedness.

In practice, the concept of social vulnerability focuses on identifying individuals or groups characterised by a reduced capacity to respond to a natural disaster (De Marchi and Scolobig 2012), who may therefore suffer greater losses and will be less likely to recover from the disaster as they have less access to information, motivation and financial resources, as well as to social and organisational networks. Research in this field is based on rich tradition of sociological studies of social inequality (Kuhlicke et al. 2011). Social vulnerability is usually captured using the taxonomic approach (Wisner 2004; De Marchi and Scolobig 2012), which involves two broad sets of main 'classical' indicators, relating to demographics and to the socio-economic status (Flanagan et al. 2011; Kuhlicke et al. 2011; Yoon 2012; Thomas et al. 2013; Werg et al. 2013; Rufat et al. 2015). The indicators are designed to identify the groups more vulnerable to natural disasters and likely to suffer the most from their occurrence (Yoon 2012). The former includes the characteristics of individuals attributed to them by birth, such as gender, age, race and ethnicity, and the latter refer to social characteristics acquired by an individual through his or her actions or inaction (including education, employment, occupation, income, poverty, etc.) (Yoon 2012). Both these sets of properties of social groups describe directly 'people's vulnerability', as opposed to the properties relating to a 'place's vulnerability', which characterise the area inhabited by a community under study, such as the level of urbanisation, the economy structure, socio-economic development, the character of the building development, etc. (Cutter et al. 2000; Cutter 2010; Holand et al. 2011; Yoon 2012).

The taxonomic approach has drawn the attention of the authorities and services to social groups which need to be included in disaster risk management but were previously ignored in disaster management plans, focused mainly on the flood protection system (Wisner 2004). However, it ignores the internal diversity of vulnerable groups, e.g. not all elderly people are more likely to be affected by the consequences of a natural disaster, but more so those who are socially isolated, have poorer education and suffer from health problems (Wisner 2004; Tapsell et al. 2010). As a consequence, analyses based on taxonomic approaches may be misleading since they include people and households who cope well (false positive) in the category of vulnerable groups or, worse yet, to exclude persons or households who are truly vulnerable, but do not fit into any of the proposed categories (false negative) (De Marchi and Scolobig 2012).

Therefore, a situational approach is proposed which takes into account the dependence of social vulnerability on the position of individuals or groups (Wisner 2004). It places an emphasis on the fact that social vulnerability factors may manifest themselves in certain circumstances, for example, may depend on the type of hazard. Social vulnerability should no longer be considered as a static disposition of representatives of vulnerable groups (Werg et al. 2013), but it must also take into account its temporal variability and, to a degree, its randomness. The dynamics of social vulnerability may result from events which are difficult to predict (e.g. it may intensify during short spells of illness), but may also be subject to more typical daily, weekly or yearly cycles, when the vulnerability of a person or household decreases or increases depending on the situation and resources available at a given moment. For example, most persons can be considered more vulnerable at night, when they are asleep. Elderly persons who stay home alone on a weekday, when other family members leave for work, are more vulnerable than on weekends, when the whole family is together.

However, such an approach is much more difficult to be operationalised. Though criticised, the taxonomic approach remains in widespread use, as it is useful in initial identification of potential social vulnerability to natural disasters. More in-depth understanding of the mechanisms underlying social vulnerability is possible using qualitative methods.

Two main approaches are used in research practice: case studies of populations at risk of natural disasters and spatial modelling of social vulnerability based on census data (Rufat et al. 2015). The former uses quantitative and qualitative social study methods to investigate thoroughly selected communities exposed to natural disasters. Their purpose is to show the regularities that govern individuals, households or communities, in particular by explaining mechanisms underlying their behaviour and their limitations before, during and after a disaster. Even though individual studies cannot be a basis for drawing definitive conclusions about a particular phenomenon, meta-analysis of many such studies can lead to certain generalisations (Ford et al. 2010). The latter approach is based on quantitative spatial analyses aiming to determine the differences between administrative units in terms of potential social vulnerability and its different dimensions (Cutter 2013). The former approach may take more account of the situatedness of social vulnerability, although in most research, both case studies and spatial analyses are focused on the taxonomic approach to social vulnerability (Rufat et al. 2015).

Research of social vulnerability to natural disasters based on case studies is relatively rare compared to the more widespread and standardised—to some extent—spatial analyses. Authors of quantitative studies of geographical variations of social vulnerability rely on comparable secondary data available in public statistics. When conducting case studies, they generate primary data whose scope tends to result from narrowly defined research objectives and questions (e.g. concerning the vulnerability of specific groups, such as women, the elderly, ethnic minorities). As a result, given the small number of empirical studies and diverse research strategies, it is difficult to ensure comparability of such studies. Instead, they are used for testing hypotheses concerning the impact of drivers of social vulnerability on people's behaviour, and consequently, can contribute to improving the approaches used in spatial modelling.

However, it seems that the apparent ease of performing analyses on the basis of statistical data has made many researchers disregard the serious deficiencies inherent in the method (Oulahen et al. 2015a; Działek et al. 2016), with little use made of local studies. An overview of research based on case studies shows that most of them are conducted in countries of the global South, but usually by experts representing the global North, in collaboration with local scientists, who know local languages and culture (Pelling 1997, 1998; Birkmann and Fernando 2008; Lein 2009; Armah et al. 2010; Linnekamp et al. 2011; Paul and Routray 2011; Jordan 2012; Wang et al. 2012; Taş et al. 2013; Letsie and Grab 2015). According to Kuhlicke et al. (2011), such in-depth studies are rare in the European context, with a more profound tradition in the UK only. One exception is research by German researchers in the basin of the Mulde River (Steinführer and Kuhlicke 2007). In the United States, such type of local community-level research attracted greater attention only after the catastrophic Hurricane Katrina (Elliott and Pais 2006; Chen et al. 2007; Rufat et al. 2015), which revealed many disproportionately affected social groups.

In the course of case studies, researchers use different research methods, mainly questionnaire surveys, where respondents' answers to questions are used as indicators in quantitative analyses (e.g. Birkmann and Fernando 2008; Siegrist and Gutscher 2008; Taş et al. 2013), and less so research of purely qualitative nature (individual in-depth interviews, focus group interviews, participatory observations) (e.g. Alexander 2013; Chomsri and Sherer 2013). They usually use triangulation of quantitative and qualitative methods (e.g. Anderson-Berry 2003; Lein 2009; Armah et al. 2010; Kuhlicke et al. 2011; Paul and Routray 2011; De Marchi and Scolobig 2012; Oulahen et al. 2015b), by using the results of qualitative research to create survey questionnaires or to interpret survey results through interviews, workshops, etc. (e.g. Letsie and Grab 2015; Oulahen et al. 2015a).

## 4.2 Social Vulnerability Drivers

In the taxonomic approach, economic conditions (poverty/wealth) and demographic determinants (gender, age, race/ethnicity) are considered to be the factors that most strongly influence social vulnerability to natural disasters (Yoon 2012; Thomas et al. 2013; Werg et al. 2013; Rufat et al. 2015).

Financial resources, or more broadly, socio-economic status, which, in addition to wealth, also involves the power relations and access to knowledge and information, are considered to be factors of huge importance for disaster preparedness, the capacity to respond to disasters properly and to recover from them (Pelling 1997; Fothergill and Peek 2004; Wisner 2004; Bolin 2007; Linnekamp et al. 2011; Oulahen et al. 2015b; Rufat et al. 2015). Even though in absolute terms, losses suffered by poorer people as a result of a disaster are lower than those suffered by wealthy persons, relatively to the entire property, losses of the former may prove much greater since they are likely to represent a significant portion of their possessions. In addition, recovery may be more difficult for this group since their incomes do not put them in a position to take out insurance or save up sufficient sums to rebuild their houses (Cutter et al. 2003; Fothergill and Peek 2004). As is often the case, houses of low-income individuals may be in a poorer technical condition (higher physical vulnerability) or are more often located within areas at risk (higher exposure)—as a consequence they are more susceptible to the impact of the natural elements (Colten 2006; Brouwer et al. 2007; Finch et al. 2010; Walker and Burningham 2011; Fielding 2012; Dash 2013; Taş et al. 2013). Thus, a natural disaster can deepen socio-economic inequalities (Thywissen 2006; McCoy and Dash 2013). Homeless people are a special example of a group with low financial resources who remain unnoticed as a result of social exclusion, and their extremely difficult position should also be addressed by the authorities (Wisner 1998).

The level of economic resources is also associated with other characteristics deepening social vulnerability to natural disasters (Rufat et al. 2015), such as owning a house or flat (tenants are less likely to take protective action because they believe that the owner is responsible for doing this; Cutter et al. 2003; Tapsell et al. 2010; Wachtendorf et al. 2013; Oulahen et al. 2015b), having a car (easier self-evacuation), having a computer with Internet access (access to knowledge) and landline or mobile phone (easier transmission of alerts and communication with other residents when faced with the hazard).

As it seems, economic capital, human capital (educational attainment) and social capital (networks of social connections, social trust) combined together are particularly important. A higher level of education is usually correlated with better financial situation (Rufat et al. 2015). Educational attainment may translate into better access to information, greater knowledge of hazards, possible ways to prepare for them, and better understanding of incoming communications (Cutter et al. 2003; Fothergill and Peek 2004; Santos-Hernandez and Morrow 2013). Nowadays, the Internet is becoming an increasingly important channel for communicating information about hazards and educating people in this respect (Roth and Brönnimann 2013; Stanbrough 2013). Digital exclusion can involve deficits of both the economic capital (no computer with Internet access) and of human capital (Internet as a source of knowledge or entertainment).

Formal education does not always translate into better understanding of the mechanisms behind the natural world. Communities who live off farming can have more contact with nature, both with its natural rhythms and events of an exceptional character. In addition, rural populations tend to live longer in the same place. All this reinforces intergenerational transfer of knowledge about local hazards (McEwen et al. 2012). For this reason, population moving into a threatened area is considered more vulnerable (Clark et al. 1998; Birkmann and Fernando 2008; McAdoo et al. 2009). According to other interpretations, populations who have lived in rural areas for a long time are more vulnerable, which may be a consequence of the psychological mechanism of reducing the cognitive dissonance—neglecting the hazard as a result of being attached to one's living place (Willis et al. 2011; Działek 2013a). As regards farming communities, researchers point out that their higher vulnerability is due to economic reasons: their income is usually lower, and their livelihood is more susceptible to the forces of nature, which can destroy crops or livestock (Birkmann and Fernando 2008; Armah et al. 2010).

Researchers point that social networks are particularly important during evacuation (Adeola 2009) since they facilitate the transmission of warnings and allow contacts to be used when looking for a place to evacuate to. They are also important in the preparation phase, as they can streamline the sharing of knowledge about hazards and how to prepare for them (Murphy 2007; Harvatt et al. 2011; Działek et al. 2013b). Strong bonding social capital (understood as family, friendship and family ties), which is characteristic of more long-standing communities, is helpful in consolidating the social memory of a flood, sharing stories and experience related to the preparation for one, and responding to the disaster. Meanwhile, bridging social capital, which is represented, inter alia, by local associations, creates bonds between persons of different backgrounds, including between the established population and newcomers (Berke et al. 1993; Bolin and Stanford 1998; Hawkins and Maurer 2010). Resources of both forms of social capital may facilitate taking joint protective action and enhancing involvement in bottom-up initiatives (Dynes 2002; Adger 2003; Nakagawa and Shaw 2004; Murphy 2007; Działek et al. 2013a). However, some observations indicate that under certain conditions, social networks, especially those of the bonding type, can cause exclusion, adding to the vulnerability of already vulnerable groups (Kaniasty and Norris 1995; Pelling 1998; Aldrich 2011; Chomsri and Sherer 2013). Having access to a more extensive social network, one going beyond the local community is often associated with the socio-economic status, and thus increases inequality resulting from income, education and power (Cutter et al. 2003; Elliott et al. 2010).

Of the demographic and cultural determinants that are most likely to add to social vulnerability to natural disasters, the most frequently mentioned drivers include ethnicity or race, gender and age. These qualities are often treated as indicators of vulnerable groups, although their vulnerability often results from deficits of the economic, human or social capital, which was discussed above.

Increased susceptibility of ethnic minorities is often attributable to their limited financial resources, which is connected with the fact that immigrants tend to have worse paid jobs. In addition, their reduced resilience may result from their being positioned lower in the power relations, language barriers restricting access to information, as well as cultural determinants specific for some groups. Minorities may have difficulty understanding information about hazards communicated in the language of the dominant group—this applies both to the preparation for the hazard and the warning and evacuation stage, as well as post-disaster support and recovery. On the one hand, they can be victims of marginalisation and stereotypes of community members and institutions responsible for risk management, and on the other, being mistrustful towards the authorities, they may ignore official warnings, limit themselves to information from their families and neighbours or refrain from applying for assistance in the wake of the disaster (e.g. if they reside in a country illegally) (Cutter et al. 2003; Wang et al. 2012; Dash 2013).

Cultural factors are very important for the nature of relationships between genders and characteristics attributed to them, as well as expectations, behaviours, stereotypes and roles associated with genders. Theoretical considerations and empirical findings do not provide clear answers as to the role of gender in coping with natural disasters so that it could be concluded whether it is women or men that should be considered more vulnerable (Kuhlicke et al. 2011; Tobin-Gurley and Enarson 2013; Rufat et al. 2015). That women are more vulnerable may be implied by the different—depending on society—inequalities in earnings, access to education, level of security, participation in public life, in the power structures and in decision-making. On account of lower incomes and greater responsibility for family care, women can be less capable of coping with crises and their consequences (Cutter et al. 2003). In addition, women can fall prey to violence in the wake of natural disasters. Another important fact is that women have a small share in disaster risk management within the typically masculinised power structures of the authorities and crisis management services (Enarson et al. 2007; Tobin-Gurley and Enarson 2013). At the same time, women can play a greater role in creating informal neighbourly ties and can foster bottom-up activities (Steinführer and Kuhlicke 2007).

There are also situations where men can be more susceptible to natural disasters, which is related to the prevailing stereotypes of masculinity. Women, who are considered responsible for care over the family, are more likely to follow recommendations of the emergency services and proceed to evacuation, while men, who assume responsibility for the property, remain at home despite having been ordered to leave the place. Another factor is greater risk tolerance (Gustafson 1998; Finucane et al. 2000), which is reflected by an increased proportion of young men among flood victims (Ashley and Ashley 2008; Lowe et al. 2013). Greater risk tolerance can also cause men to ignore risks and fail to prepare for a flood (Tobin-Gurley and Enarson 2013).

Age is another important driver of social vulnerability to natural disasters (Peek 2013). As a rule, with age the resources, thanks to which individuals are capable of facing natural disasters increase, because their knowledge, experience, status, wealth and extent of social networking grow. Two age groups are characterised by increased social vulnerability: very young and elderly persons (Cutter et al. 2003; Birkmann and Fernando 2008; Ashley and Ashley 2008; Kuhlicke et al. 2011; Peek 2013). Children are more likely to be casualties of sudden natural disasters, such as earthquakes, tsunamis and flash floods. In the case of less violent events, such as hurricanes and heat waves, they are less exposed since parents are more likely to evacuate than people with no children (Peek 2013). Regardless of how violent an event is, children may suffer from emotional stress and health problems related, for example, to limited access to medical care after a natural disaster. However, during the preparatory period, it can be assumed that families with children should be better motivated to take precautions to protect their offspring. Children as such are identified as an important factor strengthening local ties, and they can also stimulate the involvement of adults in information exchange and flood education, thanks to school curricula, as well as during common leisure activities (Walker et al. 2012). However, children are rarely treated as addressees of communications about hazards, because it is assumed that their parents are responsible for them. Yet, many children are separated from their parents for most of the week, for example, while at school, being alone at home, playing in the backyard or in extreme cases, if they are homeless (Peek 2013). As a result, consideration should be given to educational activities about local hazards addressed to the youngest residents, which may also prove beneficial for adults in the long run.

Elderly people are disproportionately more likely to be victims of natural disasters irrespective of their character (Jonkman et al. 2009; Peek 2013). Age-related problems (limited mobility, hearing and sight disorders, attention, memory, perception and decision-making problems) can be a challenge when risks are communicated, warnings about a hazard are sent and evacuation is underway, as well as at the aid distribution and recovery stage, during which the problems can—for various reasons—cause elderly persons to be excluded when aid is distributed. Consequently, all this can affect the flood preparedness among these groups. In addition, their increased vulnerability may be due to financial constraints, possibly lower educational attainment, social isolation reducing social support, limited access to disaster-related information, including digital exclusion, which is becoming increasingly relevant in the context of the growing popularity of online and mobile communications. Elderly people can be also less likely to leave the hazard area on account of their strong attachment to their place of residence or uncertainty about whether or not anyone will care for them, whether their medicines will be available, etc. (Peek 2013; Chau et al. 2014; Meyer 2017). On the balance, the vulnerability of the group is mitigated by their greater life experience, resulting from their residing in the same area for a longer period, which often means experience of earlier floods.

The presence of children and the elderly is also associated with other social vulnerability drivers which are treated as separate characteristics in analyses of the phenomenon. Poorer health, including disability, limited mobility or chronic illnesses, are identified as one such driver, and may affect not only households with senior citizens (Davis et al. 2013; Chau et al. 2014; Twigg 2014; Rufat et al. 2015). Often, households of post-working age persons consist of one or two persons, whose resources are limited for them to cope with flood preparedness (Cutter et al. 2003;

Tapsell et al. 2010; Wachtendorf et al. 2013; Chau et al. 2014). This group also includes households of single parents or singles. Similarly, larger households are considered more vulnerable because they can be 'demographically burdened' with persons at the pre-working and post-working age, and thus are likely to have less financial resources.

## 4.3  Social Vulnerability as a Factor of Flood Preparedness

Residents of the flood-threatened areas in southern Poland both assessed their subjective sense of flood preparedness and declared actions undertaken for that purpose. The assessment of the level of flood preparedness reflects the perceived flood risk and the motivation to take protective action. A vast majority of those surveyed stated that they had been completely unprepared for the floods which had affected them. As a rule, experiencing a flood is a motivator to take preventive action. Indeed, there was an increase in the percentage of households considering themselves prepared for a flood, even though most of the residents remained pessimistic about their capacity to face another flood (Fig. 4.1).

The subjective sense of flood preparedness is partly correlated with the overall assessment of the potential for reducing flood losses (Fig. 4.2). Most of those who do not believe in the effectiveness of actions aimed at reducing such damage feel being inadequately prepared for the disaster. But also half of the respondents who believe that reducing flood damage is possible to feel insufficiently prepared for floods, which may imply an increased social vulnerability on their part.

Of the households surveyed, only one-third declare using measures to protect their houses or surroundings against floods. Interestingly, the fear of a disaster does not motivate people into taking efforts in this regard. However, the importance of the belief in the possibility of reducing flood losses for improving preparedness has been confirmed (Table 4.1).

Even though households use individual flood protections, a large proportion of them still do not feel prepared for flooding. Nearly, half of the households which

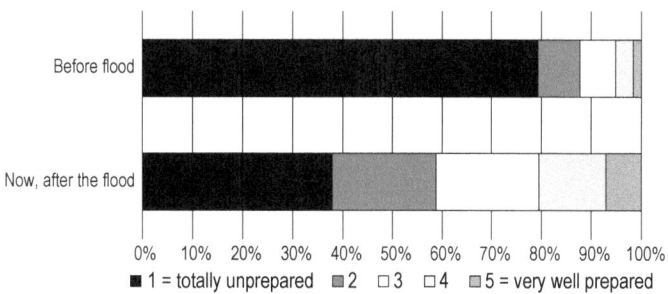

**Fig. 4.1** Self-assessment of the flood preparedness in households. *Source* Own study

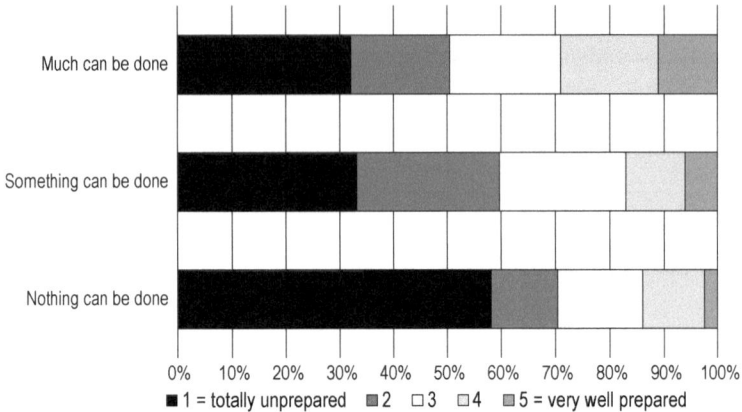

**Fig. 4.2** Potential to mitigate flood losses versus self-assessed flood preparedness. *Source* Own study

take such actions believe that they would not cope with a disaster of a similar scale. The opposite belief is expressed by every third respondent.

Social vulnerability of households associated with demographic, social, cultural and economic factors should be reflected in increased or decreased scope of protective actions taken (Zahran et al. 2008; Hossain 2015): the higher the level of social vulnerability, the lower the expected flood preparedness.

The relationships between the characteristics considered as determinants of social vulnerability (demographic factors and socio-economic status) and the aggregate flood preparedness index were analysed (Table 4.2). The index takes values from 0 to 7 points, where 0 denotes no action taken by a household and 7—declaration of all the types of action. Consequently, index values of 0 and 1 were considered as low flood preparedness, 2 and 3 as average, 4 and 5 as high, and 6 and 7 as very high.

Also, the impact of control variables not associated with social vulnerability, related to past experience of floods among the households surveyed, their perception of the flood risk and the possibility of reducing flood losses was analysed.

Among the households differing in terms of the age structure of their members, the low activity of those consisting exclusively of elderly persons is notable (Table 4.3). This confirms increased social vulnerability of this age group when faced with flood hazard. However, households consisting of people of different age groups were not found to be less prepared, which implies that their being 'burdened demographically' has no impact on their activity. Households consisting of working-age persons only and those with working-age persons who have children prove to be slightly better prepared.

Almost all the households with people aged 65 plus surveyed consisted of one or two persons (for the other groups, every fifth household represented this category). Retirement and disability benefits are the main source of income for senior citizens' households in a vast majority of the cases. In this age group, the percentage of house-

**Table 4.1**   Households that take protective action depending on their assessment of the flood risk and on the potential to reduce flood losses

| | | Share of households whose members | | |
|---|---|---|---|---|
| | | Took protective action individually | Took joint protective action with other residents | Referred to the authorities for a protective action individually or jointly with neighbours |
| Assessment of a flood likelihood in the coming few years | Will not happen | 34.1 | 15.9 | 30.7 |
| | Neither will, nor will not | 31.7 | 19.2 | 39.2 |
| | Will happen | 38.6 | 22.3 | 52.3 |
| | No answer | 31.8 | 6.8 | 34.1 |
| Assessment of the potential to reduce flood losses | Nothing can be done | 17.8 | 10.9 | 34.9 |
| | Partially can be done | 38.4 | 18.4 | 40.8 |
| | Much can be done | 44.2 | 24.3 | 58.2 |

*Source* Own study

**Table 4.2**   Specification of the flood preparedness index

| Indicators from survey | Answer | Score |
|---|---|---|
| Self-assessment of being prepared for a flood | High or very high | 1 |
| Individual flood mitigation behaviour | Yes | 1 |
| Number of flood mitigation activities selected from a set list of activities | 1–2 | 1 |
| | 3 or more | 2 |
| Possession of flood insurance | Yes | 1 |
| Collective flood mitigation behaviour with neighbours | Yes | 1 |
| Has contacted local authorities regarding flood hazards | Yes | 1 |

*Source* Own study

**Table 4.3** Relationship between flood preparedness and age structure in households

| Age structure of household members | | | N | Flood preparedness (% of households) | | | | Mean flood preparedness index | |
|---|---|---|---|---|---|---|---|---|---|
| 0–18 years | 19–64 years | 65 years of more | | Low | Average | High | Very high | In pts | Difference* |
| | x | | 171 | 20.5 | 40.4 | 32.2 | 7.0 | 3.04 | +0.28 |
| x | x | | 181 | 23.8 | 34.8 | 34.8 | 6.6 | 2.96 | +0.20 |
| x | x | x | 126 | 25.4 | 42.1 | 27.0 | 5.6 | 2.73 | −0.03 |
| | x | x | 137 | 22.6 | 46.7 | 25.5 | 5.1 | 2.82 | +0.06 |
| | | x | 107 | 48.6 | 30.8 | 15.9 | 4.7 | 1.91 | −0.85 |

*Difference relative to the mean for all household surveyed ($N = 726$)

*Source* Own study

holds which find their income level unsatisfactory is higher than among the other groups. In addition, they slightly more often declare that their financial situation has been getting worse. On average, the inhabitants of such households are also characterised by lower level of education and almost twice less frequent association activity. They also demonstrate a higher level of mistrust—nearly half of them (compared with one-third among the other households) consider that their neighbours mind their own business, rather than commit themselves to helping others. The proportion of households with persons with disabilities and suffering from chronic diseases is also higher among the elderly. This confirms the hypothesis that, in this group of households, deficits of resilience to negative phenomena add up, resulting, in this case, in poorer flood preparedness.

Households with children have not turned out to be less prepared for floods. In addition, after analysing households with three or more children, it was found that they were generally better prepared for floods (Table 4.4). Thus, while children as such may be more vulnerable, especially when unattended, households with children are not that vulnerable. On the contrary, the survey confirms that they can be better motivated to take protective actions before a disaster to protect their offspring.

Potentially, one- and two-person households can handle floods in the least effective way, while four- and five-person households can cope with floods most successfully (Table 4.5). Of the one-person households, nearly half were classified as the least prepared ones, and only one in four was considered to be well or very well prepared. Opposite proportions were recorded among five-person households.

Flood preparedness was also investigated in the context of the gender of household members. Those where adult men or women predominated did not demonstrate either radically lower or higher preparedness (Table 4.6). It is only in the case of households consisting of women or men only that one can identify—in both cases—groups of reduced activity, especially for women. In this case, different dimensions of vulner-

**Table 4.4** Relationship between flood preparedness and presence of children in a household

| Presence of children in a household | N | Flood preparedness (% of households) | | | | Mean flood preparedness index | |
|---|---|---|---|---|---|---|---|
| | | Low | Average | High | Very high | In pts | Difference* |
| Households without children** | 361 | 24.7 | 42.1 | 26.9 | 6.4 | 2.80 | +0.04 |
| Households with children | 308 | 24.4 | 37.7 | 31.5 | 6.5 | 2.87 | +0.11 |
| Including households with more than 3 children | 37 | 18.9 | 29.7 | 40.5 | 10.8 | 3.41 | +0.65 |

*Difference relative to the mean for all household surveyed ($N = 726$)
**Households with elderly only were excluded
*Source* Own study

**Table 4.5** Relationship between flood preparedness and household size

| Household size | N | Flood preparedness (% of households) | | | | Mean flood preparedness index | |
|---|---|---|---|---|---|---|---|
| | | Low | Average | High | Very high | In pts | Difference* |
| 1 person | 68 | 45.6 | 29.4 | 19.1 | 5.9 | 2.10 | −0.66 |
| 2 persons | 155 | 29.7 | 41.3 | 23.9 | 5.2 | 2.59 | −0.17 |
| 3 persons | 123 | 27.6 | 38.2 | 28.5 | 5.7 | 2.70 | −0.06 |
| 4 persons | 149 | 19.5 | 44.3 | 30.9 | 5.4 | 2.98 | +0.22 |
| 5 persons | 105 | 21.0 | 33.3 | 36.2 | 9.5 | 3.12 | +0.36 |
| 6 persons and more | 125 | 24.8 | 41.6 | 28.0 | 5.6 | 2.79 | +0.03 |

*Difference relative to the mean for all household surveyed ($N = 726$)
*Source* Own study

**Table 4.6** Relationship between flood preparedness and the gender structure of households

| Gender structure of households | N | Flood preparedness (% of households) | | | | Mean flood preparedness index | |
|---|---|---|---|---|---|---|---|
| | | Low | Average | High | Very high | In pts | Difference* |
| More women than men (2/3 and more) | 193 | 27.5 | 32.6 | 33.7 | 6.2 | 2.80 | +0.04 |
| More men than women (2/3 and more) | 140 | 30.7 | 37.9 | 25.7 | 5.7 | 2.66 | −0.10 |
| Women only | 68 | 39.7 | 30.9 | 23.5 | 5.9 | 2.25 | −0.51 |
| Men only | 36 | 33.3 | 44.4 | 19.4 | 2.8 | 2.44 | −0.32 |

*Difference relative to the mean for all household surveyed ($N = 726$)
*Source* Own study

ability add up since they are usually small households, and among women, who live longer on average, they are usually households of elderly people. Thus, it cannot be concluded whether poorer preparedness is a result of increased vulnerability of either gender—it seems that the predominant factor here might be the scarcer resources of such households resulting from their size and advanced age of their members.

Households with disabled or chronically ill persons may be more vulnerable not only during the evacuation period but also during the preparation stage, since they may require more care and greater expenditure, limiting their capacity to prepare for a disaster. Against the background of all households, indeed, those with sick or disabled persons prove to be slightly worse prepared (Table 4.7). However, when broken down by the individual types, it can be concluded that this largely applies to households of elderly persons, where health problems intensify the vulnerability to a natural disaster. On the other hand, where elderly persons live with younger people,

**Table 4.7** Relation between flood preparedness and age structure of households and presence of disabled persons

| Age structure of household members | | | N | Household with chronically ill, disabled, or persons with limited mobility | | | |
|---|---|---|---|---|---|---|---|
| | | | | Yes | | No | |
| 0–18 years | 19–64 years | 65 years or more | | % of house-holds | Mean flood pre-paredness index in pts | % of house-holds | Mean flood pre-paredness index in pts |
| | x | | 171 | 19.9 | 2.76 (0.00) | 80.1 | 3.10 (+0.34) |
| x | x | | 181 | 17.1 | 3.00 (+0.24) | 82.9 | 2.95 (+0.19) |
| x | x | x | 126 | 49.2 | 2.82 (+0.06) | 50.8 | 2.64 (−0.12) |
| | x | x | 137 | 50.4 | 2.74 (−0.02) | 49.6 | 2.90 (+0.16) |
| | | x | 107 | 42.1 | 1.64 (−1.12) | 57.9 | 2.10 (−0.66) |
| Total | | | 726 | 33.5 | 2.59 (−0.17) | 66.5 | 2.84 (+0.08) |

*Note* In parentheses difference relative to the mean for all household surveyed ($N = 726$)
*Source* Own study

households inhabited by persons suffering from health problems are, on average, slightly better prepared than others, suggesting that—as in the case of households with children—they seem to be more active to ensure security for the less able family members.

The level of flood preparedness is pronouncedly higher for households of self-employed persons or persons employed on a full-time basis who occupy non-farming jobs than in case where farming or retirement and disability benefits are the main source of income (Table 4.8). This is attributable to income gaps. In general, the more such households are satisfied with their level of income, the higher their average level of flood preparedness (Table 4.9). Flood preparedness is correlated similarly with the evaluation of the financial situation of households—where the situation has deteriorated significantly in recent years, the level of preparedness has declined.

The financial situation of a household is a better predictor of taking protective action than evaluation of the change in the situation in recent years (Table 4.10). Households which consider their current situation to be positive, irrespective of whether it has deteriorated or improved over the last 5 years, form a group of relatively better prepared households.

The index takes into account having a car, a computer with Internet access, and landline and mobile phone by the household. It is not surprising that households

**Table 4.8** Relationship between flood preparedness and the main source of income of households

| Main source of income of households | N | Assessment of the financial situation* | Flood preparedness (% of households) | | | | Mean flood preparedness index | |
|---|---|---|---|---|---|---|---|---|
| | | | Low | Average | High | Very high | In pts | Difference** |
| Employment outside agriculture | 287 | 81.8 | 18.1 | 40.8 | 33.4 | 7.7 | 3.08 | +0.28 |
| Self-employment outside agriculture, or rental income | 73 | 87.5 | 20.5 | 34.2 | 38.4 | 6.8 | 3.15 | +0.39 |
| Farming employment | 59 | 64.4 | 39.0 | 39.0 | 18.6 | 3.4 | 2.24 | −0.52 |
| Retirement or disability pension | 295 | 63.6 | 32.9 | 38.6 | 23.4 | 5.1 | 2.50 | −0.26 |

*Share of households declaring that their current income is sufficient to meet basic needs
**Difference relative to the mean for all household surveyed ($N = 726$)
*Source* Own study

without one of the above were generally more poorly prepared (Table 4.11). However, owning the house or flat occupied by the household is more important. There were relatively few households not owning the house or flat they were residing in at the time of the study (less than 8%), which is characteristic of smaller towns in Poland. The only exceptions were the towns of Bieruń and Bogatynia, where large hard and brown coal mines are located. Non-owners are among the households who very rarely take protective action. This confirms that the group is particularly vulnerable to natural disasters because they belong to the less well-off groups, or believe that such actions should be taken by the landlord, and not by the tenant.

The second important determinant of the socio-economic status which may translate into social vulnerability, and consequently, into flood preparedness, is the level of education. The results of the study confirm this relation (Table 4.12).

Here, the resources of human capital seem to be more important than those of economic capital (Table 4.13). Households with better educated members are generally better prepared, regardless of whether they consider their financial situation to be good or bad. By contrast, households characterised by lower levels of education are additionally affected by the negative impact of their poorer financial situation.

The resources of the bonding and bridging social capital can be represented in several ways. The bonding social capital of household members may result from the time period over which they have resided in a given area. The longer they have lived there, the greater the expected network of neighbourly relations, which should be conducive to the flow of information about local threats and ways of preventing them.

**Table 4.9** Relationship between flood preparedness and the financial situation of households

| Assessment of the financial situation of households | | N | Flood preparedness (% of households) | | | | Mean flood preparedness index | |
|---|---|---|---|---|---|---|---|---|
| | | | Low | Average | High | Very high | In pts | Difference* |
| Current income is sufficient to meet basic needs | Yes | 99 | 23.2 | 40.4 | 29.3 | 7.1 | 2.96 | +0.20 |
| | Rather yes | 426 | 21.8 | 42.0 | 30.3 | 5.9 | 2.86 | +0.10 |
| | Yes—total | 525 | 22.1 | 41.7 | 30.1 | 6.1 | 2.88 | +0.12 |
| | Rather no | 120 | 38.3 | 30.0 | 25.8 | 5.8 | 2.49 | −0.27 |
| | No | 76 | 39.5 | 35.5 | 19.7 | 5.3 | 2.33 | −0.43 |
| | No—total | 196 | 38.8 | 32.1 | 23.5 | 5.6 | 2.43 | −0.26 |
| Financial situation of the household in the last 5 years | Clearly improved | 9 | 0.0 | 55.6 | 33.3 | 11.1 | 3.33 | +0.57 |
| | Improved | 87 | 20.7 | 41.4 | 29.9 | 8.0 | 3.00 | +0.24 |
| | Improved—total | 96 | 18.8 | 42.7 | 30.2 | 8.3 | 3.03 | +0.27 |
| | Not Changed | 410 | 25.9 | 39.0 | 30.5 | 4.6 | 2.74 | −0.02 |
| | Deteriorated | 161 | 26.7 | 40.4 | 25.5 | 7.5 | 2.79 | +0.03 |
| | Clearly deteriorated | 56 | 46.4 | 30.4 | 16.1 | 7.1 | 2.30 | −0.46 |
| | Deteriorated—total | 217 | 31.8 | 37.8 | 23.0 | 7.4 | 2.66 | −0.10 |

*Difference relative to the mean for all household surveyed ($N = 726$)
*Source* Own study

**Table 4.10** Mean value of the synthetic flood preparedness index depending on the financial situation of households

| | | Mean flood preparedness index in pts | | |
|---|---|---|---|---|
| | | Financial situation of the household in the last 5 years | | |
| | | Improved | Not changed | Deteriorated |
| Current income is sufficient to meet basic needs | Yes | 3.06 (+0.30) | 2.80 (+0.04) | 2.98 (+0.22) |
| | No | x | 2.49 (−0.27) | 2.40 (−0.36) |

*Note* In parentheses difference relative to the mean for all household surveyed ($N = 726$)
*Source* Own study

**Table 4.11** Relationship between flood preparedness and material situation of households

| Material situation of households | | N | Flood preparedness (% of households) | | | | Mean flood preparedness index | |
|---|---|---|---|---|---|---|---|---|
| | | | Low | Average | High | Very high | In pts | Difference* |
| Owns car | Yes | 543 | 21.2 | 41.3 | 30.8 | 6.8 | 2.97 | +0.21 |
| | No | 183 | 43.2 | 32.8 | 20.2 | 3.8 | 2.13 | −0.63 |
| Owns computer with Internet access | Yes | 539 | 22.3 | 40.6 | 31.2 | 5.9 | 2.93 | +0.17 |
| | No | 187 | 39.6 | 34.8 | 19.3 | 6.4 | 2.26 | −0.50 |
| Owns landline phone | Yes | 477 | 23.7 | 39.0 | 31.2 | 6.1 | 2.92 | +0.16 |
| | No | 249 | 32.5 | 39.4 | 22.1 | 6.0 | 2.44 | −0.32 |
| Owns mobile phone | Yes | 655 | 25.3 | 39.5 | 29.0 | 6.1 | 2.82 | +0.06 |
| | No | 71 | 39.4 | 35.2 | 19.7 | 5.6 | 2.15 | −0.61 |
| Owns house or flat | Yes | 685 | 25.0 | 39.6 | 29.2 | 6.3 | 2.83 | +0.07 |
| | No | 40 | 55.0 | 32.5 | 10.0 | 2.5 | 1.55 | −1.21 |

*Difference relative to the mean for all household surveyed ($N = 726$)
*Source* Own study

As it turns out, however, households present in an area for a relatively short time (up to 10 years) are better prepared on average (Table 4.14). This may be attributed to increased interest on the part of newcomers in the local environment shortly after their arrival.

However, there is a correlation between flood preparedness and ties between residents. Those who believe that their relationships are not good are generally more poorly prepared (Table 4.14). Less trusting households may have less developed networks of ties and thus may participate in the flow of knowledge about hazards to a limited extent. They can be also less likely to receive support, also when preparing for a potential flood.

The hypothesis regarding resources of bridging social capital, which is represented by the activity of household members belonging to the voluntary fire service and other associations and organisations, also proves to be true. Households whose

**Table 4.12** Relationship between flood preparedness and education of household members

| Level of education of household members | N | Flood preparedness (% of households) | | | | Mean flood preparedness index | |
|---|---|---|---|---|---|---|---|
| | | Low | Average | High | Very high | In pts | Difference* |
| Predominance of adults with secondary and higher education (2/3 and more) | 291 | 15.8 | 40.5 | 36.1 | 7.6 | 3.23 | +0.47 |
| Predominance of adults with basic vocational education, lower secondary education or lower education (2/3 and more) | 279 | 40.9 | 35.1 | 21.1 | 2.9 | 2.17 | −0.59 |

*Difference relative to the mean for all household surveyed ($N = 726$)
*Source* Own study

**Table 4.13** Mean value of the synthetic flood preparedness index depending on the level of education and on the financial situation of households

| | | Mean flood preparedness index in pts | |
|---|---|---|---|
| | | Predominant level of education of household members | |
| | | Secondary and higher education | Basic vocational education, lower secondary education or lower education |
| Current income is sufficient to meet basic needs | Yes | 3.26 (+0.50) | 2.37 (−0.39) |
| | No | 3.21 (+0.45) | 1.82 (−0.96) |

*Note* In parentheses difference relative to the mean for all household surveyed ($N = 726$)
*Source* Own study

members belonged to one of the two types of organisation proved to be better prepared (Table 4.15).

Interesting results were obtained by examining the relationship between flood preparedness and the combination of social, economic and human capital (Table 4.16). Households considering their incomes to be poor and with lower educational attainment are made even more vulnerable if their members are not involved in the activities of local associations or the voluntary fire service. Households of better educated per-

**Table 4.14** Relationship between flood preparedness and bonding social capital of households

| Indicators of bonding social capital of households | | N | Flood preparedness (% of households) | | | | Mean flood preparedness index | |
|---|---|---|---|---|---|---|---|---|
| | | | Low | Average | High | Very high | In pts | Difference* |
| Length of residence in an area | Less than 10 years | 37 | 18.9 | 27.0 | 45.9 | 8.1 | 3.35 | +0.59 |
| | 10–29 years | 90 | 24.4 | 37.8 | 32.2 | 5.6 | 2.88 | +0.12 |
| | 30–49 years | 112 | 28.6 | 42.0 | 25.9 | 3.6 | 2.57 | −0.19 |
| | 50 years and more | 485 | 27.4 | 39.6 | 26.6 | 6.4 | 2.73 | −0.03 |
| Evaluation of relations between residents | Everyone minds their own business | 259 | 33.2 | 40.5 | 21.6 | 4.6 | 2.41 | −0.35 |
| | Neither so, nor so | 136 | 19.9 | 36.8 | 35.3 | 8.1 | 3.04 | +0.28 |
| | People support and help each other | 322 | 24.8 | 37.9 | 31.1 | 6.2 | 2.92 | +0.16 |

*Difference relative to the mean for all household surveyed ($N = 726$)
*Source* Own study

**Table 4.15** Relationship between flood preparedness and bridging social capital of households

| Indicators of bridging social capital of households | | N | Flood preparedness (% of households) | | | | Mean flood preparedness index | |
|---|---|---|---|---|---|---|---|---|
| | | | Low | Average | High | Very high | In pts | Difference* |
| Households with members of volunteer firefighter brigade | Yes | 118 | 20.3 | 35.6 | 34.7 | 9.3 | 3.15 | +0.39 |
| | No | 607 | 28.0 | 39.7 | 26.9 | 5.4 | 2.68 | −0.08 |
| Households with members of associations | Yes | 129 | 14.7 | 37.2 | 34.9 | 13.2 | 3.43 | +0.67 |
| | No | 595 | 29.4 | 39.5 | 26.7 | 4.4 | 2.61 | −0.15 |

*Difference relative to the mean for all household surveyed ($N = 726$)
*Source* Own study

**Table 4.16** Mean value of the synthetic flood preparedness index depending on membership in social organisations, the financial situation and level of education

| | | Mean flood preparedness index in pts | | | |
|---|---|---|---|---|---|
| | | Households with members of associations | | Households with members of volunteer firefighter brigade | |
| | | Yes | No | Yes | No |
| Current income is sufficient to meet basic needs | Yes | 3.39 (+0.63) | 2.75 (−0.01) | 3.28 (+0.52) | 2.80 (+0.04) |
| | No | 3.52 (+0.76) | 2.27 (−0.49) | 2.71 (−0.05) | 2.38 (−0.38) |
| Predominance of people with secondary and higher education | | 3.77 (+1.01) | 3.08 (+0.32) | 3.10 (+0.34) | 3.26 (+0.50) |
| Predominance of people with basic vocational education, lower secondary education or lower education | | 2.79 (+0.03) | 2.09 (−0.67) | 2.98 (+0.22) | 2.02 (−0.74) |

*Note* In parentheses difference relative to the mean for all household surveyed ($N = 726$)
*Source* Own study

sons involved in the activities of associations prove to be the most resilient to the potential hazard. A careful analysis of these relationships indicates that the level of human and social capital has a stronger influence on preparedness than economic capital. For example, households declaring that their incomes are not sufficient for satisfying their basic needs, but whose members engage in associative activity, prove to be equally well prepared as those declaring good financial standing. Also, households whose members have lower educational attainment, but are active in the local community, prove to be prepared on an average level (when they are association members), or are above the average (when they are members of the voluntary fire service).

According to many studies, prior experience of a natural disaster is a very strong predictor of greater risk awareness, and consequently, of taking of protective actions to reduce losses during subsequent events (Smith 2001; Siegrist and Gutscher 2008; Wachinger and Renn 2010; Harvatt et al. 2011; Działek 2013b; Rufat et al. 2015). In particular, it is stressed that the scale and frequency of losses as a result of a disaster are relevant: people affected by one catastrophic event are less likely to take protective actions than those affected by a disaster repeatedly, since the former seem to treat such a disaster as exceptional, one that will not happen again (Tobin and Burrel 1997; Renn 2008).

Generally, concerns about the possibility of a disastrous event motivate intensified preparatory action, even though such action may be constrained by psychological mechanisms, e.g. in a situation when the likelihood of losses is judged as low or acceptable, or where those at risk do not know how to counter a potential hazard, which can make them deny its existence (Raaijmakers et al. 2008; Willis et al. 2011). Persons with internal locus of control believe that damage can be avoided by their assuming personal responsibility for action, while those with external locus of control consider natural disasters to be divine punishment or fate, which results in a sense of hopelessness and lack of preparation, despite prior flood experience (McClure et al. 1999; Bell et al. 2005; Lin et al. 2008; Siegrist and Gutscher 2008; Armaş and Avram 2009).

The households surveyed differ in terms of the frequency of flood losses, the period which has elapsed since the last major disaster and the magnitude of the losses (Table 4.17). The time factor does not differentiate the willingness to take protective actions, as opposed to the frequency of being affected by floods which cause losses. It has been confirmed that those repeatedly affected by floods are more motivated to take up protective action than those who have experienced only one big flood. The amount of losses also explains greater commitment to flood preparedness, with households who have suffered the greatest losses showing above average activity.

Also, the fear of flood reoccurring in a given area motivates people to take precautions. The respondents who believed that the flood in their locality would not repeat, belonged to the group of the most poorly prepared households on average (Table 4.18). Those who were avoiding the answer to this question were not very active, which suggests that they may evade thinking about the hazard, which translates into their lack of commitment to preventing a possible event. However, of all the psychological determinants of preparedness for events threatening a household, the sense of being capable of limiting the resultant losses proves to be the most highly differentiating factor. The difference in this respect between optimists (much can be done) and pessimists (nothing can be done) is noticeable—greater than for the previously analysed correlation with demographic and socio-economic factors.

Various relations between flood preparedness and the characteristics considered to be drivers of social vulnerability were examined so far. These factors are interrelated (see Walker and Burningham 2011; Yoon 2012; Rufat et al. 2015), which requires applying linear regression analysis to determine which of them best explain the differences in flood preparedness observable (see Elliott et al. 2010; Kuhlicke et al. 2011). The synthetic flood preparedness index was adopted as the dependent variable, and the demographic and socio-economic drivers likely to influence social vulnerability were chosen as the independent variables. They were expressed as dummy variables where value 1 represented higher hypothetical social vulnerability, and consequently poorer flood preparedness. Thus, the results of regression with a negative sign for a given variable will therefore confirm the role of the respective driver in increasing the vulnerability, and thus limiting the flood preparation capacity. A positive sign means that (unlike what is indicated by theoretical proposals) the level of vulnerability decreases, and the preparedness of such a household is higher. For the control variables, the indicators were also constructed so that value 1 corresponded—in line with source literature—to higher flood preparedness. The backward elimination method was applied, starting with all the independent variables in the primary model. Following this, in each step, one variable with the lowest

**Table 4.17** Relationship between flood preparedness and prior flood experience

| Prior flood experience | | N | Flood preparedness (% of households) | | | | Mean flood preparedness index | |
|---|---|---|---|---|---|---|---|---|
| | | | Low | Average | High | Very high | In pts | Difference* |
| How many times have they suffered losses as a result of floods or local flooding over the last dozen or so years? | Once | 342 | 30.4 | 40.4 | 24.9 | 4.4 | 2.55 | −0.19 |
| | More than once | 384 | 23.4 | 38.0 | 31.0 | 7.6 | 2.94 | +0.18 |
| When was the flood that caused the greatest damage? | 2008 or earlier | 175 | 27.4 | 39.4 | 26.9 | 6.3 | 2.74 | −0.02 |
| | 2009 or later | 550 | 26.5 | 38.9 | 28.5 | 6.0 | 2.76 | 0.00 |
| Assessment of the size of damage caused by the flood | Very large | 337 | 23.4 | 38.6 | 30.3 | 7.7 | 2.93 | +0.17 |
| | Large | 177 | 27.7 | 42.4 | 26.0 | 4.0 | 2.61 | −0.15 |
| | Average | 159 | 31.4 | 35.8 | 26.4 | 6.3 | 2.65 | −0.11 |
| | Small | 53 | 30.2 | 41.5 | 26.4 | 1.9 | 2.47 | −0.29 |

*Difference relative to the mean for all household surveyed ($N = 726$)
*Source* Own study

significance level was eliminated until a set of variables with a significance level below the assumed threshold, i.e. 0.1, was reached.

Before proceeding to the interpretation of the results, it should be recalled that they merely indicate that specific characteristics of households are likely to be regarded as increasing or decreasing their vulnerability to floods—it does not automatically mean that each holding with a given characteristic belongs to 'vulnerable groups' (Kuhlicke et al. 2011; De Marchi and Scolobig 2012).

The regression analysis for households affected by floods demonstrates that the discrepancies between their preparedness for a potential hazard are best explained by the determinants of social vulnerability describing their socio-economic status and the control variables (Table 4.19). Most of the demographic characteristics prove to be statistically insignificant. The analysis has only confirmed a generally increased

**Table 4.18** Relationship between flood preparedness and flood risk perception

| Flood risk perception | | N | Flood preparedness (% of households) | | | | Mean flood preparedness index | |
|---|---|---|---|---|---|---|---|---|
| | | | Low | Average | High | Very high | In pts | Difference* |
| Assessment of a flood likelihood in the coming few years | 1 = will not happen | 48 | 43.8 | 41.7 | 10.4 | 4.2 | 2.04 | −0.72 |
| | 2 | 40 | 30.0 | 37.5 | 27.5 | 5.0 | 2.60 | −0.16 |
| | 3 | 120 | 25.0 | 46.7 | 28.3 | | 2.51 | −0.25 |
| | 4 | 147 | 23.1 | 38.8 | 33.3 | 4.8 | 2.87 | +0.11 |
| | 5 = will happen | 283 | 23.0 | 35.0 | 31.4 | 10.6 | 3.11 | +0.35 |
| | Brak odpowiedzi | 88 | 36.4 | 42.0 | 18.2 | 3.4 | 2.23 | −0.53 |
| Assessment of the potential to reduce flood losses | Nothing can be done | 129 | 51.2 | 32.6 | 14.7 | 1.6 | 1.86 | −0.90 |
| | Partially can be done | 331 | 25.1 | 45.3 | 25.1 | 4.5 | 2.63 | −0.13 |
| | Much can be done | 251 | 15.5 | 33.9 | 39.8 | 10.8 | 3.43 | +0.67 |

*Difference relative to the mean for all household surveyed ($N = 726$)
*Source* Own study

vulnerability to natural disasters of households with elderly persons, who, given their poorer access to resources, are already more poorly prepared. It has also been demonstrated that households with many children are among those slightly better prepared on average, as they are probably more motivated to protect their offspring. The variables describing the gender structure, the size of households and the presence of people with disabilities cannot be considered statistically significant.

A majority of the variables describing the resources of the households surveyed explain flood preparedness in a significant way. A specific and, at the same time, surprising exception is the indicator of the subjective evaluation of the financial situation of a household, which seems to be of key significance as a barrier to investing in protection. Neither the presence of an unemployed person in the household nor the lack of a computer with Internet access is significant. All the statistically significant variables, except for one, indicate interrelations corresponding to the hypotheses made. Only households of newcomers, unlike what was expected, are generally better prepared for another flood.

The determinants of social vulnerability with the highest power explaining reduced flood preparedness are as follows: lower level of education of household

**Table 4.19**   Regression analysis on flood preparedness index

| | | Households | Standardised ß-coefficients |
|---|---|---|---|
| Social vulnerability drivers | Demographic characteristics | With more than 3 children | 0.077 |
| | | With seniors only | −0.111 |
| | | With female majority | |
| | | With male majority | |
| | | Small (1 or 2 persons) | |
| | | Large (6 persons or more) | |
| | | With handicapped or chronically ill person | |
| | Socio-economic status | With negative assessment of own financial situation | |
| | | With unemployed | |
| | | With main source of income from farming | −0.071 |
| | | Not owning house or flat | −0.092 |
| | | Not owning computer with Internet access | |
| | | With lower educational status | −0.177 |
| | | With negative assessment of local social ties | −0.091 |
| | | Moved to the area less than 10 years ago | 0.067 |
| | | With no membership in volunteer firefighter brigade | −0.064 |
| | | With no membership in associations | −0.118 |
| Control variables | Prior experience | Flooded several times (vs. only once) | 0.087 |
| | | With very high losses during last flood | 0.107 |
| | Risk perception | That believe there will be flood over the next several years | 0.092 |
| | | That believe much versus nothing can be done to reduce flood losses | 0.210 |
| $R$ | | | 0.499 |
| Adjusted $R^2$ | | | 0.235 |

*Note* Table shows the standardised ß-coefficients of final model of the stepwise linear regression with backward elimination with a significance lower than 0.1
*Source* Own study

members, non-involvement in association activity, elderly households, not owning the house or flat and negative evaluation of local societal ties. On the other hand, the sense that reducing losses is possible was one of the control variables with the strongest positive effect on flood preparedness. This characteristic has the greatest explanatory power of all the characteristics used in the model. The other three control characteristics also provide statistically significant information (of a similar value) to the model explaining flood preparedness.

# References

Adeola FO (2009) Katrina cataclysm: does duration of residency and prior experience affect impacts, evacuation, and adaptation behavior among survivors? Environ Behav 41(4):459–489

Adger WN (2003) Social capital, collective action, and adaptation to climate change. Econ Geogr 79(4):387–404

Aldrich DP (2011) The externalities of strong social capital: post-tsunami recovery in southeast India. J Civ Soc 7(1):81–99

Alexander M (2013) Constructions of flood vulnerability across an etic-emic spectrum. PhD thesis, Middlesex University. http://eprints.mdx.ac.uk/13530 (15.05.2016)

Anderson-Berry LJ (2003) Community vulnerability to tropical cyclones: Cairns, 1996–2000. Nat Haz 30:209–232

Armah FA, Yawson DO, Yengoh GT, Odoi JO, Afrifa EKA (2010) Impact of floods on livelihoods and vulnerability of natural resource dependent communities in Northern Ghana. Water 2(2):120–139

Armaş I, Avram E (2009) Perception of flood risk in Danube Delta, Romania. Nat Hazards 50:269–287

Ashley ST, Ashley WS (2008) Flood fatalities in the United States. J Appl Meteorol Climatol 47(3):805–818

Bell PA, Greene T, Fisher J, Baum AS (2005) Environmental psychology. Harcourt College, Orlando

Berke P, Kartez J, Wenger D (1993) Recovery after disaster: achieving sustainable development, mitigation and equity. Disasters 17:93–109

Birkmann J (ed) (2006a) Measuring vulnerability to natural hazards: towards disaster resilient societies. United Nations University Press, Tokyo, New York, Paris

Birkmann J (2006b) Indicators and criteria for measuring vulnerability: theoretical bases and requirements. In: Birkmann J (ed) Measuring vulnerability to natural hazards: towards disaster resilient societies. United Nations University Press, Tokyo, New York, Paris, pp 55–77

Birkmann J, Fernando N (2008) Measuring revealed and emergent vulnerabilities of coastal communities to tsunami in Sri Lanka. Disasters 32(1):82–105

Birkmann J, Cardona OD, Carreño ML, Barbat AH, Pelling M, Schneiderbauer S, Kienberger S, Keiler M, Alexander D, Zeil P, Welle T (2013) Framing vulnerability, risk and societal responses: the MOVE framework. Nat Hazards 67:193–211

Blaikie P, Cannon T, Davis I, Wisner B (1994) At risk: natural hazards, people's vulnerability, and disaster. Routledge, London

Bolin B (2007) Race, class, ethnicity, and disaster vulnerability. In: Rodríguez H, Quarantelli EL, Dynes RR (eds) Handbook of disaster research. Springer, New York, pp 113–129

Bolin R, Stanford L (1998) The Northridge earthquake: community-based approaches to unmet recovery needs. Disasters 22:21–38

Brouwer R, Akter S, Brander L, Haque E (2007) Socioeconomic vulnerability and adaptation to environmental risk: a case study of climate change and flooding in Bangladesh. Risk Anal 27(2):313–326

Chau PH, Gusmano MK, Cheng JOY, Cheung SH, Woo J (2014) Social vulnerability index for the older people—Hong Kong and New York City as examples. J Urban Health Bull N Y Acad Med 91(6):1048–1064

Chen AC-C, Keith VM, Leong KJ, Airriess C, Li W, Chung K-Y, Lee C-C (2007) Hurricane Katrina: prior trauma, poverty and health among Vietnamese-American survivors. Int Nurs Rev 54:324–331

Chomsri J, Sherer P (2013) Social vulnerability and suffering of flood-affected people: case study of 2011 mega flood in Thailand. Kasetsart J (Soc Sci) 34:491–499

Clark G, Moser S, Ratick S, Dow K, Meyer W, Emani S, Jin W, Kasperson J, Kasperson R, Schwartz H (1998) Assessing the vulnerability of coastal communities to extreme storms: the case of Revere, MA, USA. Mitig Adapt Strat Glob Change 3(1):59–82

Colten CE (2006) Vulnerability and place: flat land and uneven risk in New Orleans. Am Anthropol 108(4):731–734

Cutter SL (2010) Social science perspectives on hazards and vulnerability science. In: Beer T (ed) Geophysical hazards: minimizing risk, maximizing awareness. Springer, Dordrecht, pp 17–30

Cutter SL (2013) Vulnerability. In: Bobrowsky PT (ed) Encyclopedia of natural hazards. Springer, Dordrecht, pp 1088–1090

Cutter SL, Mitchell JT, Scott MS (2000) Revealing the vulnerability of people and places: a case study of Georgetown County, South Carolina. Ann Assoc Am Geogr 90(4):713–737

Cutter SL, Boruff BJ, Shirley WL (2003) Social vulnerability to environmental hazards. Soc Sci Q 84(1):242–261

Cutter SL, Emrich CT, Webb JJ, Morath D (2009) Social vulnerability to climate variability hazards: a review of the literature. Hazards and Vulnerability Research Institute, University of South Carolina, Columbia

Dash N (2013) Race and ethnicity. In: Thomas DSK, Phillips BD, Lovekamp WE, Fothergill A (eds) Social vulnerability to disasters. CRC Press, Boca Raton, London, New York, pp 113–139

Davis EA, Hansen R, Kett M, Mincin J, Twigg J (2013) Disability. In: Thomas DSK, Phillips BD, Lovekamp WE, Fothergill A (eds) Social vulnerability to disasters. CRC Press, Boca Raton, London, New York, pp 199–234

De Marchi B, Scolobig A (2012) The views of experts and residents on social vulnerability to flash floods in an Alpine region of Italy. Disasters 36(2):316–337

Dynes R (2002) The importance of social capital in disaster response. Preliminary paper #327. University of Delaware, Disaster Research Centre

Działek J (2013a) Cognitive dissonance. In: Bobrowsky PT (ed) Encyclopedia of natural hazards. Springer, Dordrecht, pp 98–99

Działek J (2013b) Perception of natural hazards and disasters. In: Bobrowsky PT (ed) Encyclopedia of natural hazards. Springer, Dordrecht, pp 756–759

Działek J, Biernacki W, Bokwa A (2013a) Challenges to social capacity building in flood-affected areas of southern Poland. Nat Hazards Earth Syst Sci 13:2555–2566

Działek J, Biernacki W, Bokwa A (2013b) Impact of social capital on local communities' response to floods in Southern Poland. In: Neef A, Shaw R (eds) Risks and conflicts: local responses to natural disasters. Community, environment and disaster risk management, vol 14. Emerald, pp 185–205

Działek J, Biernacki W, Fiedeń Ł, Listwan-Franczak K, Franczak P (2016) Universal or context-specific social vulnerability drivers—understanding flood preparedness in southern Poland. Int J Disaster Risk Reduct 19:212–223

Elliott JR, Pais J (2006) Race, class, and Hurricane Katrina: social differences in human responses to disaster. Soc Sci Res 35:295–321

Elliott JR, Haney TJ, Sams-Abiodun P (2010) Limits to social capital: comparing network assistance in two New Orleans neighborhoods devastated by Hurricane Katrina. Sociol Q 51(4):624–648

Enarson E, Fothergill A, Peek L (2007) Gender and disaster: foundations and directions. In: Rodríguez H, Quarantelli EL, Dynes RR (eds) Handbook of disaster research. Springer, New York, pp 130–146

Fekete A (2012) Spatial disaster vulnerability and risk assessments: challenges in their quality and acceptance. Nat Hazards 61:1161–1178

Fekete A, Hufschmidt G, Kruse S (2014) Benefits and challenges of resilience and vulnerability for disaster risk management. Int J Disaster Risk Sci 5(1):3–20

Fielding JL (2012) Inequalities in exposure and awareness of flood risk in England and Wales. Disasters 36(3):477–494

Finch C, Emrich CT, Cutter SL (2010) Disaster disparities and differential recovery in New Orleans. Popul Environ 31(4):179–202

Finucane ML, Slovic P, Mertz CK, Flynn J, Satterfield TA (2000) Gender, race, and perceived risk: the 'white male' effect. Health Risk Soc 2(2):159–172

Flanagan BE, Gregory EW, Hallisey EJ, Heitgerd JL, Lewis B (2011) A social vulnerability index for disaster management. J Homel Secur Emerg Manage 8(1):1–22

Ford JF, Keskitalo ECH, Smith T, Pearce T, Berrang-Ford L, Duerden F, Smit B (2010) Case study and analogue methodologies in climate change vulnerability research. Wiley Interdiscip Rev Clim Change 1(3):374–392

Fordham M, Lovekamp WE, Thomas DSK, Phillips BD (2013) Understanding social vulnerability. In: Thomas DSK, Phillips BD, Lovekamp WE, Fothergill A (eds) Social vulnerability to disasters. CRC Press, Boca Raton, London, New York, pp 1–29

Fothergill A, Peek L (2004) Poverty and disasters in the United States: a review of recent sociological findings. Nat Hazards 32(1):89–110

Fox-Rogers L, Devitt C, O'Neill E, Brereton F, Clinch JP (2016) Is there really "nothing you can do"? Pathways to enhanced flood-risk preparedness. J Hydrol 543(B):330–343

Gall M (2007) Indices of social vulnerability to natural hazards: a comparative evaluation. PhD thesis, University of South Carolina. http://webra.cas.sc.edu/hvri/education/docs/Melanie_Gall_2007.pdf (15.05.2016)

Gustafson PE (1998) Gender differences in risk perception: theoretical and methodological perspectives. Risk Anal 18(6):805–811

Harvatt J, Petts J, Chilvers J (2011) Understanding householder responses to natural hazards: flooding and sea-level rise comparisons. J Risk Res 14(1):63–83

Hawkins RL, Maurer K (2010) Bonding, bridging and linking: how social capital operated in New Orleans following Hurricane Katrina. Br J Soc Work 40(6):1777–1793

Holand IS, Lujala P, Rød JK (2011) Social vulnerability assessment for Norway: a quantitative approach. Norsk Geografisk Tidsskrift—Norw J Geogr 65(1):1–17

Hossain MN (2015) Analysis of human vulnerability to cyclones and storm surges based on influencing physical and socioeconomic factors. Evidences from coastal Bangladesh. Int J Disaster Risk Reduct 13:66–75

Jonkman SN, Maaskant B, Boyd E, Levitan ML (2009) Loss of life caused by the flooding of New Orleans after Hurricane Katrina: analysis of the relationship between flood characteristics and mortality. Risk Anal 29(5):676–698

Jordan E (2012) Pathways to community recovery: a qualitative comparative analysis of post-disaster outcomes. PhD thesis, University of Colorado

Kaniasty K, Norris FH (1995) In search of altruistic community: patterns of social support mobilization following Hurricane Hugo. Am J Community Psychol 23:447–477

Kuhlicke C, Scolobig A, Tapsell S, Steinführer A, DeMarchi B (2011) Contextualizing social vulnerability: findings from case studies across Europe. Nat Hazards 58(2):789–810

Lein H (2009) The poorest and most vulnerable? On hazards, livelihoods and labelling of riverine communities in Bangladesh. Singap J Trop Geogr 30:98–113

Letsie MM, Grab SW (2015) Assessment of social vulnerability to natural hazards in the mountain kingdom of Lesotho. Mt Res Dev 35(2):115–125

Lin S, Shaw D, Ho M-C (2008) Why are flood and landslide victims less willing to take mitigation measures than the public? Nat Hazards 44(2):305–314

Linnekamp F, Koedam A, Baud ISA (2011) Household vulnerability to climate change: examining perceptions of households of flood risks in Georgetown and Paramaribo. Habitat Int 35:447–456

Lowe D, Ebi KL, Forsberg B (2013) Factors increasing vulnerability to health effects before, during and after floods. Int J Environ Res Public Health 10(12):7015–7067

McAdoo BG, Moore A, Baumwoll J (2009) Indigenous knowledge and the near field population response during the 2007 Solomon Islands tsunami. Nat Hazards 48(1):73–82

McClure J, Walkey F, Allen M (1999) When earthquake damage is seen as preventable: attributions, locus of control and attitudes to risk. Appl Psychol 48:239–256

McCoy B, Dash N (2013) Class. In: Thomas DSK, Phillips BD, Lovekamp WE, Fothergill A (eds) Social vulnerability to disasters. CRC Press, Boca Raton, London, New York, pp 83–112

McEwen L, Krause F, Hanson J, Jones O (2012) Flood histories, flood memories and informal flood knowledge in the development of community resilience to future flood risk. In: 11th British Hydrological Society, National Hydrology Symposium, 9–11 July 2012. http://eprints.uwe.ac.uk/20319 (15.05.2016)

Meyer MA (2017) Elderly perceptions of social capital and age-related disaster vulnerability. Disaster Med Public Health Prep 11(1):48–55

Murphy BL (2007) Locating social capital in resilient community-level emergency management. Nat Hazards 41(2):297–315

Nakagawa Y, Shaw R (2004) Social capital: a missing link to disaster recovery. Int J Mass Emerg Disasters 22(1):5–34

Oulahen G, Mortsch L, Tang K, Harford D (2015a) Unequal vulnerability to flood hazards: "ground truthing" a social vulnerability index of five municipalities in Metro Vancouver, Canada. Ann Assoc Am Geogr 105(3):473–495

Oulahen G, Shrubsole D, McBean G (2015b) Determinants of residential vulnerability to flood hazards in Metro Vancouver, Canada. Nat Hazards 78:939–956

Paul SK, Routray JK (2011) Household response to cyclone and induced surge in coastal Bangladesh: coping strategies and explanatory variables. Nat Hazards 57:477–499

Peek L (2013) Age. In: Thomas DSK, Phillips BD, Lovekamp WE, Fothergill A (eds) Social vulnerability to disasters. CRC Press, Boca Raton, London, New York, pp 167–198

Pelling M (1997) What determines vulnerability to floods: a case study in Georgetown, Guyana. Environ Urban 9(1):203–226

Pelling M (1998) Participation, social capital and vulnerability to urban flooding in Guyana. J Int Dev 10:469–486

Raaijmakers R, Krywkow J, van der Veen A (2008) Flood risk perceptions and spatial multi-criteria analysis: an exploratory research for hazard mitigation. Nat Hazards 46(3):307–322

Raška P (2015) Flood risk perception in Central-Eastern European members states of the EU: a review. Nat Hazards 79(3):2163–2179

Renn O (2008) Risk governance: coping with uncertainty in a complex world. Earthscan, London

Roth F, Brönnimann G (2013) Focal report 8: risk analysis using the internet for public risk communication. Risk and Resilience Research Group, Center for Security Studies (CSS), ETH Zürich

Rufat S, Tate E, Burton CG, Maroof AS (2015) Social vulnerability to floods: review of case studies and implications for measurement. Int J Disaster Risk Reduct 14:470–486

Santos-Hernandez JM, Morrow BH (2013) Language and literacy. In: Thomas DSK, Phillips BD, Lovekamp WE, Fothergill A (eds) Social vulnerability to disasters. CRC Press, Boca Raton, London, New York, pp 265–280

Siegrist M, Gutscher H (2008) Natural hazards and motivation for mitigation behavior: people cannot predict the affect evoked by a severe flood. Risk Anal 28(3):771–778

Smith K (2001) Environmental hazards: assessing risk and reducing disaster. Routledge, London, New York

Stanbrough L (2013) Internet, world wide web and natural hazards. In: Bobrowsky PT (ed) Encyclopedia of natural hazards. Springer, Dordrecht-New York, pp 563–565

Steinführer A, Kuhlicke C (2007) Social vulnerability and the 2002 flood: country report Germany (Mulde River). FLOODsite report T11-07-08

Tapsell S, McCarthy S, Faulkner H, Alexander M (2010) Social vulnerability and natural hazards. CapHaz-Net WP4 report. Flood Hazard Research Centre—FHRC, Middlesex University, London. http://caphaz-net.org/outcomes-results/CapHaz-Net_WP4_Social-Vulnerability2. pdf (15.05.2016)

Taş M, Taş N, Durak S, Atanur G (2013) Flood disaster vulnerability in informal settlements in Bursa, Turkey. Environ Urban 25(2):443–463

Tate E (2012) Social vulnerability indices: a comparative assessment using uncertainty and sensitivity analysis. Nat Hazards 63:325–347

Thomas DSK, Phillips BD, Lovekamp WE, Fothergill A (eds) (2013) Social vulnerability to disasters. CRC Press, Boca Raton, London, New York

Thywissen K (2006) Core terminology of disaster reduction: a comparative glossary. In: Birkmann J (ed) Measuring vulnerability to natural hazards: towards disaster resilient societies. United Nations University Press, Tokyo, New York, Paris, pp 448–496

Tobin GA, Burrel EM (1997) Natural hazards: explanation and integration. Guilford Press, New York

Tobin-Gurley J, Enarson E (2013) Gender. In: Thomas DSK, Phillips BD, Lovekamp WE, Fothergill A (eds) Social vulnerability to disasters. CRC Press, Boca Raton, London, New York, pp 139–166

Twigg J (2014) Attitude before method: disability in vulnerability and capacity assessment. Disasters 38(3):465–482

UN/ISDR (2004) Living with risk: a global review of disaster reduction initiatives, t. 1. United Nations, New York, Geneva

UN/ISDR (2012) Building resilience to disasters in Europe: connect and convince to reduce impact of vulnerability. United Nations, Geneva

Vávra J, Lapka M, Cudlínová E, Dvořáková-Líšková Z (2015) Local perception of floods in the Czech Republic and recent changes in state flood management strategies. J Flood Risk Manag. https://doi.org/10.1111/jfr3.12156

Wachinger G, Renn O (2010) Risk perception and natural hazards. CapHaz-Net WP3 report. DIALOGIK, Stuttgart. http://caphaz-net.org/outcomes-results (15.05.2016)

Wachtendorf T, Nelan MM, Blinn-Pike L (2013) Households and families. In: Thomas DSK, Phillips BD, Lovekamp WE, Fothergill A (eds) Social vulnerability to disasters. CRC Press, Boca Raton, London, New York, pp 281–310

Walker G, Burningham K (2011) Flood risk, vulnerability and environmental justice: evidence and evaluation of inequality in a UK context. Crit Soc Policy 31(2):216–240

Walker M, Whittle R, Medd W, Burningham K, Moran-Ellis J, Tapsell S (2012) 'It came up to here': learning from children flood narratives. Children's Geogr 10(2):135–150

Wang M-Z, Amati M, Thomalla F (2012) Understanding the vulnerability of migrants in Shanghai to typhoons. Nat Hazards 60:1189–1210

Werg J, Grothmann T, Schmidt P (2013) Assessing social capacity and vulnerability of private households to natural hazards—integrating psychological and governance factors. Nat Hazards Earth Syst Sci 13:1613–1628

Willis KF, Natalier K, Revie M (2011) Understanding risk, choice and amenity in an urban area at risk of flooding. Housing Stud 26(2):225–239

Wisner B (1998) Marginality and vulnerability: why the homeless of Tokyo don't 'count' in disaster preparations. Appl Geogr 18(1):25–33

Wisner B (2004) Assessment of capability and vulnerability. In: Bankoff G, Frerks G, Hilhorst D (eds) Mapping vulnerability: disasters, development and people. Earthscan, London, pp 183–193

Yoon DK (2012) Assessment of social vulnerability to natural disasters: a comparative study. Nat Hazards 63:823–843

Zahran S, Brody SD, Peacock WG, Vedlitz A, Grover H (2008) Social vulnerability and the natural and built environment: a model of flood casualties in Texas. Disasters 32(4):537–560

# Chapter 5
# Online Flood Risk Communication

In addition to the very experience of a flood remembered once the flood is over, which is discussed in Chap. 3, exchange of information during the event itself, in the aftermath, and at the time when people are no longer under the direct impact of the stimulus that a flood is, seems to be the crucial process modifying the memory of floods. Literature considers that it is crucially important, when adopting the study procedure, to distinguish between the above stages in determining the actual impact of communication on the awareness of at-risk populations (Lindell and Perry 2004; Steinführer et al. 2009).

Studies of natural disaster-related communication are linked to a growing popularity of research of the social aspects of natural disasters (Höppner et al. 2010). The behaviour of people before and during a disaster results from the risk perceived by the threatened population (Działek 2013). Risk perception and mitigation activities depend on the experience people have, and when they lack it, on their knowledge, which highlights the crucial importance of informing and educating people by the competent authorities. It is also stressed that communication should be tailored to the social vulnerability of specific groups, to their needs and limitations (Biernacki 2013).

Without support through communication, over time the picture of the area affected by the event is effaced out of the public memory, practical knowledge disappears, and as a consequence, the human cognitive process leads—as was shown in the previous chapters—to increased social vulnerability to flood. The relationship, especially in the context of communication, is an area which has been poorly researched to date (Tapsell et al. 2010).

As is demonstrated by research, the level of dialogue, two-way communication varies considerably from one part of the world to another (Chambers 2003; Delli Carpini et al. 2004; Parkins and Mitchell 2005; Stanghellini and Collentine 2008). At the same time, existing research indicates that there is a need for departing from an approach to communication based on 'instructions' issued by experts in favour of one based on dialogue and understanding of people's attitudes (Walls et al. 2004;

© The Author(s), under exclusive licence to Springer Nature Switzerland AG 2019     91
J. Działek et al., *Understanding Flood Preparedness*, SpringerBriefs in Geography,
https://doi.org/10.1007/978-3-030-04594-4_5

Höppner et al. 2012). A high awareness, which can be enhanced by communication, is considered a first step towards reducing vulnerability (Bostrom et al. 2008).

Research across the world has revealed that people seek information about natural hazards online (Roth and Brönnimann 2013; Stanbrough 2013), although often they do this only when a natural disaster is imminent (Bird et al. 2008; Lee et al. 2009). In Poland, Internet access is already widespread, also in peripheral areas, and the problem lies to a growing extent on the competences of the senders and addressees (Czerniawska 2012; Batorski 2013), which is particularly important for information on potential hazards. The survey conducted as part of the project presented here shows that 23% of the respondents preferred municipalities' websites as a source of information about floods. In addition, 7 and 7.5% of those surveyed considered other websites with local news and the social media, respectively, as an important source of information. The latter are an increasingly important means of virtual human communication, using text, sound, image or film to convey the message. They offer new ways of social participation in risk management such as citizen reporting, digital volunteerism and crowdsourced crisis mapping (Alexander 2014; Palen and Hughes 2017). One of such media is YouTube, the most popular video-sharing website (Cheng et al. 2008), which creates huge opportunities for reaching and influencing wide audiences (Simonsen 2011; YouTube 2017).

One advantage of web-based communication is that it allows contents to be presented in various forms (text, images, videos), and accordingly, it is suitable for diverse forms of activity (reading, writing, watching, listening, talking) (Komac et al. 2010). In addition, the contents are available 24 h a day. The diversity of forms of online communication permits targeting selected groups (e.g. groups vulnerable to a hazard), which are non-numerous from the perspective of traditional media (e.g. people with disabilities, ethnic minorities). The Internet and social media are therefore a more open and accessible medium, more useful for two-way real-time communication (Bricout 2010; Höppner et al. 2012; Roth and Brönnimann 2013; Alexander 2014; Houston et al. 2015), both between residents and between residents and institutions responsible for risk management.

However, the use of social media and the Internet as a tool of dialogue between citizens and risk managers encounters a lot of obstacles. Generators of disaster social media content could be various representing national, regional and local governments, emergency services, media, other organisations and last but not least individuals (Houston et al. 2015). They can generate information of different qualities and veracities. The Internet content is more and more frequently criticised as a source of unverified information, where gossip or deliberate misinformation can circulate freely (Roth and Brönnimann 2013). During natural disasters, social media serve as a place where people can share their emotional reactions, rather than engage in rational discussion—however, they can impact social change in longer perspective (Neubaum et al. 2014; Al-Saggaf and Simmons 2015). Hence, websites of trusted public institutions can play an important role both at the central level (description of countrywide hazards and general rules of conduct) and the local level (specificities of local risks and tailored advice). However, the working hours of officials act as a significant limitation since demand for information is round the clock.

Research also shows that during disasters, emergency services are trained to use standardised protocols, which are not suitable for less predictable online exchanges via social media and make responders uncertain how to react. At the same time, there are usually not enough members of staff to support online communication and handle large amounts of information and numbers of queries during such events (Hughes et al. 2014). The role of disaster social media is, however, much wider and includes exchange of information not only during but especially before and directly after an event (Houston et al. 2015).

In this chapter, we concentrate on two different types of online risk communication: first, we assess the accessibility and usefulness of the content generated by different categories of public institutions (national and regional, then local) to be used during pre-event phase, and second, we are attempting to better understand the patterns of content recorded by individuals during floods and uploaded afterwards in social media, and relate it to their potential to build social memory of floods and to risk education.

## 5.1  Information and Educational Content Published on Websites of National and Regional Institutions Engaged in Flood Prevention Activities

Public authorities are obliged in various ways to develop communication with the public and to inform actively not only ordinary people but also social and economic actors, of their actions. There are a number of elements which stimulate these activities, but the most important ones include the following:

1. Duties to share or provide information to certain social groups.
2. Obligations ensuing from the objectives of an institution or its activities formulated in a general way by law.
3. Constitutional obligations of institutions towards the general public.

Institutions' websites are a key means facilitating communication and sharing information with any entity. Naturally, communication, information sharing or active transfer of information do not have any formal effect—they do not oblige anyone to act, but can highly motivate the recipients of such information to engage in action or fulfil their respective duties. As a consequence, analysing the key characteristics of such stakeholders can be important for understanding existing situation and can help in designing instruments relevant for such institutions with a view to improving their communication with other stakeholders.

In fact, Polish legislation does not define the powers of institutions dealing with floods in the field of information, communication or provision of information. It can only be assumed that some of their activities, such as public consultations or warning residents, require defining the principles of communication, putting them in place and monitoring their effectiveness. However, existing legislation does not contain any suggestions on how to do this.

It is assumed that the following groups are recipients of information in the risk management communication process:

– children and youth,
– adult residents and potential investors,
– local authorities,
– central government administration.

The analysis was based on assessing the difference between the competences of a unit in respect of communication with individual groups of customers and the actual actions the unit takes in this regard using its websites.

Both the competences and actions were defined on a scale of 1–3, defined as follows:

1. Low competences—providing basic information about the priorities, policies or strategic plans in areas falling within the remit of the institution. At this level, neither direct contact with a given target audience nor supporting the institutions which may be responsible for such contact are required.
2. Medium competences—supporting institutions which have indirect influence on selected target groups, such as units of local government, non-governmental organisations, teacher associations, etc. These competencies also involve establishing standards for communication by financing materials of crucial importance for the target groups, such as textbooks, recommendations, etc., and making them available on their websites. No active direct contact is required.
3. High competences—direct contact with specific target groups, understood as clear indication on the institution's website that it makes materials (guides, manuals, leaflets), or services, such as training services, smartphone applications, etc. available on its website. Having an indirect influence on institutions which have direct contact with target groups also falls within the scope of standard competences at this level.

Taking into account the above definitions of the qualifiers, it is proposed that the institutions under investigation be divided into the following three groups:

1. State administration units of ministerial level (Ministry of the Interior and Administration, Ministry of the Environment, the Government Centre for Security—Rządowe Centrum Bezpieczeństwa) should play an important role in communication with other units of central and local government, especially those at national and regional levels, mainly as regards applicable law, national policies and priorities, as well as central guidelines and recommendations, as these institutions are responsible for initiating them. This role will be less pronounced with regard to educating residents and children—the above units should support lower level authorities, with a focus on promoting the key policies resulting from the policy of the state, e.g. by funding pilot activities, methodological approaches, and, for example, annual nationwide competitions for the best flood protection actions delivered by local governments, the state administration and NGOs. They could also finance key guidance and textbooks with information and knowledge useful in education and provision of information.

2. Specialised central administration units (the National Water Management Authority, the Institute of Meteorology and Water Management, the Chief Sanitary Inspector of the State, the National Headquarters of the State Fire Service) should strongly support the activities of specific groups of units (of the state administration, and local government) in implementing applicable legal requirements and methodological recommendations resulting from flood risk mitigation plans and policies. This includes such solutions as providing information about flood hazards (system of flood hazard and risk maps), flood risk management plans and the warning system. Because some plans prepared at this level require coordination with independent units (e.g. flood risk management plans), the task of these institutions is to support joint action between the administration, local governments and NGOs working at different levels. The above units should also support school education, as there are no specialised units addressing the problem.

3. Specialised regional administration units (Regional Water Management Authorities). In addition to managing infrastructure owned by the state, their duties include planning activities (water management and flood risk management plans for the water regions) and issuing administrative decisions, for example, consenting to exemptions from bans on development within flood areas, agreeing local land-use plans, etc. Using their local branches, they can maintain direct contact with the units of local government at risk and influence them by cooperating with the local and regional media, and non-governmental organisations.

## 5.1.1 Assessment of the Activity of Selected Institutions

As regards actions, their assessment was made according to identical criteria as those described above, except that it did not focus exclusively on the flood-related activity of the units, but their competences in general. Thus, the evaluations address the general way in which institutions act as regards communication, rather than their flood risk management activities.

The chapter below presents brief descriptions of the activities taken by each institution with regard to the groups mentioned above.

**Ministry of the Environment**
The Ministry is responsible for many institutions dealing with water management and flood risk management. It plays a key role in preparing legislative solutions and strategic approaches for this area.

The information on the Ministry's website covers many aspects, such as the principles of lawmaking, strategies and plans, registers and records, results of public consultation and many more. However, it lacks specific information about the policies of the state in individual areas, except for a policy for adaptation to climate change as well as the Central Environmental Policy for 2009–2012 with an Outlook for 2016. However, the latter is so deeply hidden on the Ministry's website that it is easier to find it with an external search engine than directly on the institution's website. An example of the way the Ministry communicates with target audiences is the law allowing landowners to cut down trees without requesting relevant authorities for a permit. The issue is commented widely on the Ministry's website, but the comments appeared only after the media had intervened following the entry of the controversial law in force. Beforehand, the Ministry had not notified its intention to change the law in this area and had not provided any reasons for the revision of legislation.

Generally, information about floods, flood risks and the government's policy to mitigate the consequences of floods is unavailable on the Ministry's website. If it is there, it is very short, limited to notices consisting of several sentences redirecting the readers to websites of the National Water Management Authority.

The Ministry appreciates environmental education as an essential aspect of its activities, entrusted to the Department of Education and Communication, which has been active for many years now. However, the great number of materials, useful mainly in schools, does not include educational materials concerning floods, either for adults or for school children.

Evaluation of activity: central administration—1, local government—0, adults—2, school children—2.

**Ministry of the Interior and Administration**
The Ministry of the Interior and Administration is an institution enjoying wide competences as regards the standards applicable to crisis management and (the financing) of flood recovery, and thus, naturally enough, units of the central government and local governments are the most obvious targets of its communications.

The website offers multiple options to be used by the public administration, and in particular by local governments. The Ministry runs an e-learning platform the purpose of which is to prepare better the staff of local government for their work. It also runs the administracja.mswia.gov.pl website, where it publishes huge amounts of information addressed to local governments. Actually, it lacks information about floods, apart from a well-written article about the legal bases on which local governments can apply for flood recovery co-financing.

Communication with citizens is not dedicated much space on the website—the various e-learning courses for local governments do not cover information or communication with local communities. Paradoxically enough, the website offers training in the protection of information.

The Ministry has also released several articles for the general public, most of which are one-page advice on how to behave in different situations, for example, during floods, storms, strong winds, etc. Longer releases with advice address terrorist attacks or car, bus, train, or air travel. An interesting article was published in 2016—'Safe Water 2016', which is a colour children's book with simple tips on how to behave by the water, on a beach, to be safe.

Evaluation of activity: central administration—1, local government—3, adults—3, children—2.

**The Government Centre for Security**

The goal of the Government Centre for Security (Rządowe Centrum Bez-pieczeństwa—RCB) is to 'optimise and unify the perception of hazards by individual ministries, and thus improve the capacities for handling difficult situations by relevant services and public administrations'. As a result, the RCB's main partners are central government and regional government institutions.

The RCB publishes regular (daily, weekly and quarterly) reports on crisis situations recorded over a given period, as well as forecasts of such events. It also prepares twice a year CIIP Focus newsletters on ICT protection. The RCB website also contains a wealth of information and links to legal acts governing crisis management activities.

The Government Centre for Security also publishes guides for the public on how to behave in various crisis situations, including 'Bezpieczny urlop' (A Safe Holiday), 'Jak przygotować się do zimy' (How to Get Ready for Winter), 'Jak przygotować się na podróż zimą samochodem' (How to Get Ready for a Winter Car Trip), 'Bezpieczeństwo na lodzie' (Safety on Ice), 'Jak przygotować się do powodzi' ('Getting Ready for a Flood'), 'Jak przygotować się na burzę' (Be Ready for a Thunder Storm), 'Jak przygotować się do wichury' (How to Get Ready for a Gale), 'Zagrożenia okresowe występujące w Polsce' (Seasonal Risks in Poland), 'Powódź—w obliczu zagrożenia' (In the Face of Flood Hazards), 'Zalecenia dla ludności zamieszkałej na terenach osuwiskowych' (Recommendations for People Living in Landslide Areas), 'Jak postępować po powodzi' (How to Behave after a Flood) and 'Wulkany w Europie' (Volcanoes in Europe). Unfortunately, after recent redesign of the website, accessing these materials has become more difficult. They can only be accessed via the site map.

Evaluation of activity: central administration—3, local government—3, adults—3, children—0.

**The National Water Management Authority**

The President of the National Water Management Authority is the central authority of the government administration responsible for water management, in particular, for matters related to the use of waters. Being responsible for drafting water management plans and flood risk management plans, the President defines the policies for action in these areas, not only those delivered by the public administration but also by other actors, including local government and central institutions.

The National Water Management Authority has several intranet sites. However, apart from the main website of the institution (kzgw.gov.pl), it is difficult to tell what the purpose of the powodz.gov.pl or geoportal.kzgw.gov.pl sites is. Not to mention the information posted on the websites of ISOK (IT System for the Protection of the Country against Extreme Hazards) or IMGW (Institute of Meteorology and Water Management), which are supervised by the National Water Management Authority.

The above websites contain information which may be of interest to many entities, but the form in which they are presented and their layout within the website may cause difficulty in understanding or even accessing them. This is exemplified by flood hazard maps and flood risk maps.

Although the institution releases materials of interest to many users, it does not do much to make their use easier. It does not even try to present the policy underlying the plans in plain language, publishes plans written in specialised language, and what is more, texts referred to as 'summaries in non-technical language' are equally difficult to understand as the plans themselves. There is no specific information addressed to local governments, although they are important partners obliged to implement specific actions, both in connection with water management plans and flood risk management plans.

One of the KZGW's websites, namely, powodz.gov.pl, includes teaching materials: a family flood plan, a guide for teachers, multimedia presentations for teachers and a brochure on how to organise flood education in a municipality.

Evaluation of activity: central administration—1, local government—1, adults—1, children—2.

**The Institute of Meteorology and Water Management**
The purposes of the Institute of Meteorology and Water Management (IMGW) include maintaining the observation and measurement network, conducting observations and measurements, collection of data and preparation of forecasts and warnings related to weather events and water levels in rivers. By law, the Institute is obliged to inform and alert central and regional institutions.

A special tool has been prepared—IMGW Monitor (Monitor IMGW)—which makes information and forecasts of the Institute of Meteorology and Water Management available to institutions lower in the administrative hierarchy, e.g. county governments. As a result, the same data, products and functionalities are made available both to the staff of the Institute and external users named in the Institute's statutes, such as emergency services, public institutions, the military and the fire service. The IMGW Monitor provides access to meteorological and hydrological data, as well as information about water levels in reservoirs. In addition, the Monitor publishes warnings and reports prepared for the individual groups of users.

Even though the Institute publishes considerable amounts of information, the website lacks descriptions on how and for what purposes the information can be used.

The Institute also offers a tool which actively warns municipal emergency services against hazards, but it can only be used for a fee. Similar commercial offers are available for businesses and the general public.

The website of the Institute also contains a range of educational materials, including the following leaflets: the 'Hydrological Cycle' (Cykl hydrologiczny), 'Why and How Should We Care about Climate' (Dlaczego i jak powinniśmy dbać o klimat), 'About the Institute' (O instytucie), 'The Meteorological Garden' (Ogródek meteorologiczny), 'Weather Forecast' (Prognoza pogody)and 'Hazards' (Zagrożenia). Even though the covers of the leaflets suggest that they are addressed to children, their content is difficult to understand and is only suitable for youths. In addition to the above materials, the pogodynka.pl website also contains educational materials entitled 'Lightning Hunter' (Łowca piorunów), 'Little Meteorologist' (Mały meteorolog), 'Storm Buster' (Pogromca burz), 'Cloud Expert' (Znawca chmur) and

'Weather Expert' (Meteo expert). They are simple quizzes on topics related to meteorology.

Evaluation of activity: central administration—2, local government—2, adults—1, children—3.

The results of the analysis show a natural, competence-based division of institutions into two groups: those which are mainly preoccupied with prevention as part of their flood risk management, and those chiefly responsible for flood response. The information published on the websites of the two groups differs completely. The former group, which includes the National Water Management Authority and Regional Water Management Authorities, focuses on issues related to the following two elements of the flood risk: hazard and exposure. For the most part, the information covered include that in the flood hazard and flood risk maps published by the National Water Management Authority and in the flood hazard studies prepared by the individual Regional Water Management Authorities (RZGW). In the second group of institutions, the publications mainly concern the vulnerability of communities at risk, and include, in the first place, guidance documents prepared by the Ministry of the Interior and Administration, the Government Centre for Security and the Headquarters of the State Fire Service.

Information released by the former group concerns the flood risk and buildings and structures within floodplain areas, but little advice is available on how to mitigate the associated hazards. By contrast, institutions in the second group publish little information about the vulnerability of areas or the different types of buildings and infrastructure, and instead release tips on how to prepare for responding to the flood hazard, and what to look out for during and after a flood. In short, in the former case, the problem is defined, but there is no advice, and in the latter, the problem is not identified, but advice is given.

Probably, this is, at least in part, a consequence of the flood risk mitigation philosophy still prevalent in Poland: the state is responsible for the prevention by protecting communities at risk with flood banks and reservoirs, while the response is entrusted to residents themselves. Certainly, another factor which determines what information is published by institutions online is the lack of a tradition of interinstitutional cooperation based on common goals, rather than on legislative requirements. As a result, even though they have a common goal, the institutions do not cooperate with one another, which is evidenced by the absence of links to contents published on the websites of other institutions.

## 5.2 Online Visibility of Information About Scientific Developments and Research

This chapter presents examples of flood-related research activities carried out in Poland in 2010–2017. Naturally, this is not an exhaustive list—given the dispersed information about research conducted by the various academic and scientific centres, compiling a full list is not feasible. All the projects presented below have been funded by national funds allocated for basic research. As can be seen, there is a noticeable activity on the part of Polish academics representing various scientific disciplines, who focus on a range of flood-related aspects.

Unfortunately, it must be added that in all the cases presented here, the potential applicability of the research is not given a priority focus, which follows from the philosophy of the National Science Centre. This is quite common in the Polish system for funding science, as a consequence of which, given its level of detail, the specific, often difficult language, information collected by researchers hardly ever reach the institutions that can use them in a practical way, implement the proposed solutions, or share the results of research with a wide range of potential target groups. Information about research conducted can be found on the Internet. The studies cover a range of disciplines, from literature through management to geophysics. Unfortunately, research reports are rarely published and available in full versions.

However, there are several examples of cooperation between such institutions as Regional Water Management Authorities, the Institute of Meteorology and Water Management, the Government Centre for Security and teams of researchers to create guidance for public use and available free of charge via the websites of the institutions. One example is a guide released by the Institute of Meteorology and Water Management on how to create and use local monitoring and warning systems, or a document prepared by the Government Centre for Security describing, in an accessible language, the steps to be followed by citizens during a flood. The Rural Development Foundation has prepared a guide describing typical problems faced during a flood by farmers and containing advice on how to behave in the face of it, which is addressed predominantly to rural residents.

Such cooperation between practitioners and theoreticians is likely to be most beneficial in terms of the preparation of contents capable of reaching potentially vulnerable persons (often online).

## 5.3   Degree to Which School Textbooks are Supported by Online Contents

Schools at lower secondary and upper secondary levels, i.e. the formal education system, are among the key institutions which create the foundations for the knowledge of future citizens. Undeniably, school is a major source of information about the world for young people, but, already in the course of formal education, and especially after it is completed, the resource of such information is greatly modified as a result of experience and information that form part of human knowledge.

To date, the school subject 'geography' as included in the core curriculum of Polish lower secondary schools has not covered the term 'flood'. Even though the term flood is present in every fourth textbook, it has never represented a separate topic. The situation is similar for upper secondary schools, where the term 'flood' has never been dedicated a separate school lesson, but instead, has been present across a very broad range of topics, most in the context of global processes or hazards. Occasionally, it also mentioned in the context of the role of the media in disseminating scientific knowledge or the use of GIS technology.

One-quarter of the textbooks for secondary schools analysed include examples of floods from Poland, with the texts illustrated with photographs and diagrams. However, none of the textbooks contains references to addresses of websites as sources of flood-related information accessible to young people.

'Education for Security' introduces the topic of floods in a completely different way. All the textbooks available on the publishing market include contents related to floods and procedures to be followed in order to face the threat. The textbook that contains the richest flood-related content includes an exhaustive description of the actions to be taken to check if a given area is at risk of being flooded. It also informs what to do to prepare one's home and relatives for a flood in the best way. The book also describes, in accessible language, the optimal scenario of actions to be taken during a flood and immediately after it. The textbook even includes practical tips on how to prepare home-made disinfectants. The text ends with a set of questions and problems motivating readers to diagnose the problem of a flood at their place of residence, and a list of websites (the only case!) where you can find additional flood-related information.

The obvious conclusion is that the problem of floods in school textbooks is not supported by references to online sources of knowledge created by other Polish institutions.

## 5.4  Dissemination of Flood Risk Information via Municipalities' Websites

At the outset, it should be emphasised that this analysis does not address the question of whether municipalities undertake the activities listed here, but whether information about such actions is available to the municipalities' residents and visitors via their official websites. It must be kept in mind that they are not the only information channel, and the contents in question can reach local communities by other means.

Access to materials which could contribute to raising residents' awareness of the flood hazard is limited (Table 5.1). Most frequently (on the website of every fifth municipality), one can find descriptions of security measures taken in the municipality, and information about historical floods. Every sixth website provides information on how to prepare for floods and, mostly quite general, information for vulnerable groups (e.g. children, the elderly, persons with disabilities). Data derived from the monitoring of water levels within local catchment areas or maps illustrating areas at risk of floods are very rarely published. Municipal documents related to crisis management, such as emergency management plans and operational flood management plans, are equally rarely published (websites even lack information that such

**Table 5.1** Availability of flood preparedness information on websites of municipalities

| Region | Number of municipalities | Share of municipalities' websites containing information on | | | | | | | | | | |
| | | Protective actions in the municipality | Historical floods | How to prepare house-holds/family for a flood | How to prepare vulnerable groups | Water level monitoring | Flood risk areas | Assembly points | Evacuation routes | The crisis management plan | The operational flood management plan |
|---|---|---|---|---|---|---|---|---|---|---|---|
| Dolnośląskie | 117 | 19.7 | 9.4 | 12.0 | 16.2 | 0.9 | 0.0 | 0.0 | 0.0 | 0.0 | 0.0 |
| Małopolskie | 163 | 19.6 | 25.2 | 16.0 | 12.3 | 3.7 | 1.8 | 0.0 | 0.0 | 3.0 | 1.2 |
| Podkarpackie | 64 | 18.8 | 21.9 | 23.4 | 26.6 | 3.1 | 3.1 | 0.0 | 0.0 | 0.0 | 0.0 |
| Total | 344 | 19.5 | 19.2 | 16.0 | 16.3 | 2.6 | 1.5 | 0.0 | 0.0 | 1.5 | 0.6 |

*Source* Own study

plans exist and can be accessed at the municipal office). None of the websites under analysis contains information about the assembly points and evacuation routes for residents of areas at risk.

Information on protective actions taken in municipalities mainly concern the construction or remodelling of flood banks, and less frequently, the draining of ditches, training for firefighters, or cooperation between communities located within the same catchment area. In general, it includes information about the activity of municipalities, municipal institutions or other public institutions. In isolated cases, reference is made to initiatives undertaken jointly with residents (e.g. a network of 'flood leaders' in Wrocław, which is to improve the flow of information during a flood). Municipalities rarely decide to publish such information systematically on a regularly updated webpage or in a separate section of the website.

While materials on protective actions are available to a similar extent across the regions analysed, information about historical floods is more frequently published on the websites of municipalities in the Małopolska and Podkarpacie regions (every fourth, fifth website), and relatively rarely on the websites of the municipalities of the Dolnośląskie (Lower Silesia) region. Perhaps, the memory of the 1997 flood, which did such huge damage to Lower Silesia, has become so dim that administrators no longer publish any information about it. Sometimes, reference is made to the flood as a historical event in the form of a brief mention in the timelines of municipalities. Perhaps, this is caused by the scarcity or lack of suitable quality photographs of the flood. The development of digital photography which has taken place since has made it possible for such events to be more widely covered, especially in 2010. However, the photographs and video recordings are very seldom accompanied by a commentary on what they relate to. Material about previous flood is usually placed in the 'latest news' sections of the websites just after the most recent flood event, or in such sections as 'Photogallery', where, after some time, it tends to disappear amidst the other events or other images. Thus, in most cases, material about previous floods does not fulfil the task of solidifying local memory of a flood or building educational messages on its basis.

Contents with tips on how to prepare for a flood and behave when it comes and in the wake of it are most often found on the websites of the Podkarpackie Region. These recommendations also include contents related to vulnerable groups—children, the elderly and persons with disabilities. There are no separate webpages addressing the situation of the groups in broader terms.

Most websites with information for people at risk of floods contain a standard set of communications, most of which are derived from two sources—a leaflet prepared jointly by the Ministry of the Interior and Administration, the State Fire Service, and the Civil Defence, and a more comprehensive brochure released by the Government Centre for Security. Often, information from the above sources is published in the form of a redesigned own leaflet, or a separate webpage. Other useful feature, that is, the 'Family Flood Plan', prepared by the National Water Management Authority, proves very poorly promoted. Such general contents are very rarely accompanied by information about the local context of flood hazards, e.g. the specific flood-related conditions within a given area (e.g. in the mountains, lowlands, concerning the breach

**Table 5.2** Availability of information on how to behave in the wake of a flood on municipalities' websites

| Region | Number of municipalities | Share of municipalities' websites containing | | | |
|---|---|---|---|---|---|
| | | Tips on how to behave after a flood | Administrative procedures | Forms and applications | Addresses and contact details of institutions |
| Dolnośląskie | 117 | 8.5 | 11.1 | 5.1 | 0.0 |
| Małopolskie | 163 | 14.7 | 18.4 | 6.7 | 9.2 |
| Podkarpackie | 64 | 21.9 | 18.8 | 9.4 | 1.6 |
| Total | 344 | 14.0 | 16.0 | 6.7 | 4.7 |

*Source* Own study

of levees), which require individual approaches to the preparation and protection of households.

Maps of flood-prone areas are rarely available on municipalities' websites. We only assessed the websites which communicated such information on separate webpages dedicated to hazards, and not, for example, as part of the land-use plans. Few municipalities presenting such information have chosen different visualisation methods: from simple (sometimes poorly legible) maps to more advanced map browsers. Few municipalities publish on their websites links to local precipitation and flood-level monitoring systems.

The range of information on actions to be taken in the wake of a flood is also limited (Table 5.2). Standard advice on how to behave after a flood, for example, how to secure affected property, is provided in the guides discussed above. Such advice rarely comes with local information, such as contact details of local institutions and services which can assist in natural disaster recovery. Communications with advice on how to proceed in the face of a flood are published by nearly all the websites analysed, and can be found directly on the home page or by searching the pages, which can be considered a positive situation. However, many websites lack detailed procedures for affected persons (e.g. farmers or business owners)—such procedures are prepared on a one-off basis, in connection with a concrete event.

Communicating information on how the local community will be warned of imminent danger is an important aspect of information about flood risks. Our analysis indicates that municipalities' websites are the most commonly used channel for warning people of an imminent natural disaster (Table 5.3), with more than half of the municipalities included in the study publishing meteorological and hydrological warnings. Much fewer websites maintained by authorities remind the public of traditional emergency warning methods by means of alarm signals sounded by sirens. The system of warning sirens, originally devised for civil defence purposes, focuses on man-made threats (air raid, contamination alerts), and less so on natural hazards.

Text message systems offered by private providers are an increasingly popular means of informing community residents of hazards. The contents of such text mes-

**Table 5.3** Availability of information on how population is warned of floods on municipalities' websites

| Region | Number of municipalities | Share of municipalities' websites with information about population warning methods | | |
|---|---|---|---|---|
| | | Sirens | SMS or mobile applications | Internet |
| Dolnośląskie | 117 | 8.5 | 41.9 | 61.5 |
| Małopolskie | 163 | 13.5 | 14.1 | 50.9 |
| Podkarpackie | 64 | 20.3 | 9.4 | 50.0 |
| Total | 344 | 13.1 | 22.7 | 54.4 |

*Source* Own study

sages are drafted by representatives of the municipality, and do not only include warnings, but also refer to important events in the municipality, such as council meetings, or cultural and sport events. Municipality residents who wish to receive such messages must register with the system. It is doubtful whether such systems fully meet their goals. At the same time, information about such systems is most readily available on the websites studied: in most cases, such information is provided in the form of permanent website sections, and more than half of them contain additional graphical elements that illustrate how the system works. The good visibility of information about this population warning method results from actions taken by software vendors, which want the service to be used by as many people as possible, which is a measure for evaluation of its success by the municipalities, which pay for such systems. The lack of a coherent single mobile phone warning system should be considered a serious shortcoming. As of 2017, a countrywide warning system (referred to officially as the National Warning System), which was supposed to transmit information to television screens and to users' phones, has not been successfully implemented. Once it is completed, the need for local text message systems should be reconsidered so as to avoid communicating risks from two different sources.

Websites of municipalities rarely redirect to other web portals offering in-depth knowledge on hazards. Every third of the websites analysed include references to various private weather services which can provide information about potential hazards, such as showers or thunderstorms. Links to the more extensive 'Pogodynka' portal, maintained by the Institute of Meteorology and Water Management (IMGW), are only available on every eighth website, providing not only weather information but also meteorological and hydrological warnings. Links to institutions and services dealing with hazards (water management authorities, the fire service, the Police or the Sanitary Inspectorate) are rarely published. Links to non-governmental organisations addressing natural hazards are missing, too.

Unfortunately, it is also difficult to find on municipalities' websites direct contact details of the staff responsible for crisis management in the municipalities. Such information, sometimes after a long search, was found on every third website. The municipalities which provide such information on separate crisis management sub-pages are to be considered exemplary in this respect. In such a situation, it is not surprising that the possibility of two-way communication between municipal authorities and residents via websites is strongly limited.

In conclusion, the range of information available on municipalities' websites is limited—they offer general information prepared by central offices, with no local information added in most cases. Survey conducted among representatives of municipal emergency services have shown that the human resources of the services are limited. Often, the issues discussed here are dealt with by individual employees, who have many other responsibilities in addition to their crisis management duties. Municipalities treat their websites as a showpiece intended to create their positive image among residents, investors and tourists. Thus, there is much room for improvement of the situation related to communicating flood risks via websites by making the related messages more visible and legible so that they successfully provide all the necessary information about local hazards, and advice on how to prepare for them, respond and behave in the wake of them.

## 5.5   Videos Posted on YouTube—A Virtual, Public 'Archive' of Local Floods

It must be noted that 83% of the YouTube videos analysed for the purposes of the present study and 92% of their viewings concern the 2010 flood. The films do not show the 1997 'Millennium Flood', which can be attributed to low availability of footage dating 20 years back, the popularity of recording technologies at the time, and the fact that YouTube itself was created nearly a decade later (Burgess and Green 2009).

A vast majority of the videos (Table 5.4) was User-Generated Content (UGC), i.e. footage created by ordinary Internet users who do not produce professional films or journalist materials (Cha et al. 2007; Simonsen 2011). There were few features produced by the regional (8%), local (5%) or nationwide (2%) media, but they have failed to engage the viewers—3/5 of the videos had no comments from YT users.

More than 2/5 of all the films do not show people's actions at all. When people do appear in the recordings, they are mostly witnesses watching the actions of the services and rescuers. Flood victims taking action to protect their households, removing damage or using the assistance of rescue services appear less frequently in the videos, and crisis management teams, politicians or experts are hardly ever present in them. The flood is shown from the perspective of a witness or flood victim using the help, with no engagement on the part of specialists other than the emergency services (Table 5.5).

A vast majority (more than 4/5) of the footage was added within the first month of the event, mostly within the first several days or at times illustrating the key moments during the flood. The films focused on showing flooded areas and the losses and

**Table 5.4**  Users who posted the videos studied

| Users who posted the videos | Number of videos | Share of videos |
|---|---|---|
| User-generated content | 120 | 82.8 |
| Local media | 8 | 5.5 |
| Regional media | 11 | 7.6 |
| Nationwide media | 3 | 2.1 |
| Non-governmental organisations Services (the military, the fire service, the police) Representatives of local, regional, central administration Government agencies (Institute of Meteorology and Water Management, National Water Management Authority) | 3 | 2.1 |

*Source* Own study

**Table 5.5**  Categories of people appearing in the videos

| People appearing in the videos | Number of videos | Share of videos |
|---|---|---|
| Witnesses | 58 | 40.0 |
| Rescuers (the army, the police, medical services) | 54 | 37.2 |
| Flood victims | 41 | 28.3 |
| Representatives of local, regional, central authorities | 5 | 3.4 |
| Media representatives | 5 | 3.4 |
| Crisis management teams | 3 | 2.1 |
| Non-governmental organisations | 3 | 2.1 |
| Politicians | 1 | 0.7 |
| Experts (e.g. a professor, employee of the Regional Water Management Authority not directly engaged in the action) | – | – |
| No people | 64 | 44.1 |

*Source* Own study

**Table 5.6**  Form of the videos studied

| Forms of the videos | Number of videos | Share of videos |
|---|---|---|
| Background sound | 108 | 74.5 |
| Single shot | 88 | 60.7 |
| Many shots | 41 | 28.3 |
| Music | 33 | 22.8 |
| Still photographs | 24 | 16.6 |
| Text | 15 | 10.3 |
| Spoken commentary | 11 | 7.6 |

*Source* Own study

devastation just after the water had receded. Even films added 1 year after the flood include, for the most part, scenes of swollen river or flooded houses.

Most of the videos posted are single-shot films, with background sound (Table 5.6). A decided majority of the films are low-quality footage, made with a mobile phone, with no subsequent quality adjustment or editing. A small proportion of the posted materials are slide shows consisting of photographs and shots taken during the flood. On the basis of the sample analysed, it can be concluded that if persons posting their videos do decide to make their message more attractive, they usually do so by adding background music, and they tend to use spoken or text commentary less frequently.

The most common type of shot in the films is a view of the river and its overflow areas, with households and road infrastructure in the background (Table 5.7). The impression is that most of the recordings were made spontaneously, to capture a moment the authors considered to be important or worth sharing with wider audiences. The quality of the recording or communicating a broader message or adding a comment or another scene seem to be of secondary importance.

As it turns out, YouTube visitors tend to view most of the videos within 1 month after they are added. Some films have been viewed with the same intensity since they were added, and some—representing a similar percentage—have been accessed at various frequencies from the time of posting.[1] The latter, i.e. those enjoying varied interest over time, form a particularly interesting group since nearly every fifth film analysed proves to display a specific viewing trend (Fig. 5.1). Viewers tend to return to videos during a potential or actual flood hazard.[2]

The example of the Bogatynia video shows that viewers can return to films even several years after a flood. They tend to play certain videos once a year, in the same

---

[1] On account of the privacy policy applied by YouTube, videos of some content providers were not analysed for the number of views over time since such data was unavailable.

[2] YouTube provides only the total number of viewers, so it is not possible to separate the share of local and regional viewers from the share of viewers coming from other regions. We suppose that the share of viewers coming from the affected communities is relatively high as they use videos to refresh their memories of these events which are significant for them. However, this issue requires further studies.

**Table 5.7**  Places depicted in videos

| Places depicted in videos | Number of videos | Share of videos |
|---|---|---|
| River, overflow area | 111 | 76.6 |
| Road and rail infrastructure | 106 | 73.1 |
| Houses and farm buildings | 65 | 44.8 |
| Hydrotechnical infrastructure (levees, dam, reservoir) | 54 | 37.2 |
| Retail points (shops, banks, other services) | 30 | 20.7 |
| Buildings and facilities available to the public (school, playground, town hall, church and others) | 28 | 19.3 |
| Natural areas and farmland | 23 | 15.9 |
| Interiors of flooded buildings | 16 | 11.0 |
| Production facilities | 11 | 7.6 |

*Source* Own study

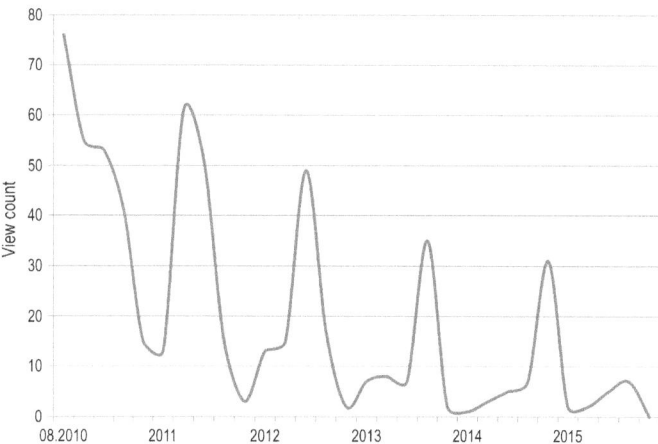

**Fig. 5.1** The number of views over time as exemplified by the video *Bogatynia powódź ul. Kościuszki. Source* Own study based on https://www.youtube.com/watch?v=vQ6S6IheTeY (accessed: 11 March 2017)

months which experienced the flood shown in the video, even if there is no flood hazard. The number of views also increases significantly when actual flood hazard reappears in a given area. One example is Tarnobrzeg, where, in May 2014, a flood alert was announced, which was followed by a high increase in the number of views of recordings depicting the previous flood (Fig. 5.2).

The videos could be divided into three types (Simonsen 2011): personal/emotional, reporting/observing and explanatory/educational. A vast majority

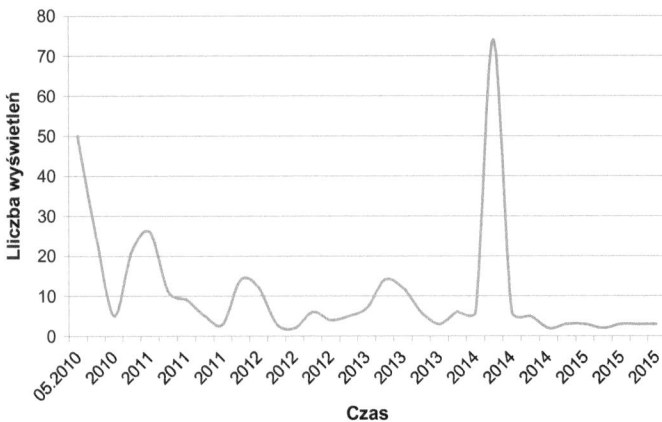

**Fig. 5.2** The number of views over time as exemplified by the video '*Strzelec' Tarnobrzeg (powódź 2010)*. *Source* Own study based on: https://www.youtube.com/watch?v=vQ6S6IheTeY (accessed: 11 March 2017)

of the films under study are of the reporting type, showing the events from an onlooker's point of view, with no emotional attitude to the situation, and do not contain any explanatory contents. The results of this observation correspond to the outcomes of a study on how various types of the social media were used during the 2011 flood in the state of Victoria, Australia (Charlwood 2012) and during natural disasters in Saudi Arabia (Al-Saggaf and Simmons 2015).

Thanks to the possibilities offered by the Internet, including the freedom of communication resulting from the nature of the social media, ordinary people witnessing a flood have been given a new information sharing method, which was still unknown a dozen or so years ago. The character of floods motivates people to share photographs and videos depicting them, raising viewers' interest, especially at times when they experience a flood and later on, when they face potential or real hazard. People capture images of a flood, creating an online record which can be accessed by entering the appropriate phrase in the search tool. Also, the media, which are subject to an increasing convergence trend and are ever more present online, contribute to the creation of such a record. Unfortunately, the contents published include little material prepared by social organisations, public services (the army, the fire service, the Police), representatives of (local, regional, central) public administration or government agencies (IMGW, KZGW). Materials prepared by them could also arouse interest among viewers and play a vital educational or informative role, complementing materials prepared by non-institutional users. Similar shortcomings were observed in YouTube coverage of the Ebola virus epidemic in West Africa (Basch et al. 2015).

Certainly, YouTube is just one example of 'intermediaries' in providing information, referred to as 'social media'. This category includes blogs, microblogs, social networking sites, forums, sharing audio, photo and video files (Balana 2012). They are all characterised by interactive communication, in which the content of the message is exchanged between private individuals, organisations and all other Internet users, regardless of their legal form or location in the real world.

The development of social media varies from country to country, but is generally dynamic in most environments and therefore all aggregate statistics are quickly becoming obsolete. Online resources are growing so fast that they are now a force to be reckoned with around the world. Nowadays, citizens can be seen as a powerful, self-organising and collectively intelligent force. For this reason, the inclusion of social media in existing risk management structures is inevitable. Social networks offer two-way communication meaning that they can include both official and unofficial information. They can be used to interact with the public and to monitor society's fears. They have significantly increased the range, size and speed of global information exchange. However, there is also a risk related to the propagation via social media of false or inaccurate information that can have potentially immense consequences. It seems that reduction of this possible danger is the biggest challenge that emergency services will have to face in the near future.

# References

Alexander DE (2014) Social media in disaster risk reduction and crisis management. Sci Eng Ethics 20(3):717–733

Al-Saggaf Y, Simmons P (2015) Social media in Saudi Arabia: exploring its use during two natural disasters. Technol Forecast Soc Chang 95:3–15

Balana CD (2012) Social media: major tool in disaster response. Inquirer Technology. http://technology.inquirer.net/12167/social-media-major-tool-in-disaster-response (30.08.2018)

Basch CH, Basch CE, Ruggles KV, Hammonds R (2015) Coverage of the Ebola virus disease epidemic on YouTube. Disaster Med Public Health Prep 9:531–535

Batorski D (2013) Polacy wobec technologii cyfrowych—uwarunkowania dostępności i sposobów korzystania. In: Diagnoza społeczna 2013. Warunki i jakość życia Polaków. Raport. Contemp Econ 7:317–341

Biernacki W (2013) Mass media and natural disasters. In: Bobrowsky P (ed) Encyclopedia of natural hazards. Springer, Dordrecht, pp 655–657

Bird D, Roberts MJ, Dominey-Howes D (2008) Usage of an early warning and information system web-site for real-time seismicity in Iceland. Nat Hazards 47:75–94

Bostrom A, French SP, Gottlieb SJ (2008) Risk assessment, modelling and decision support. Springer, New York

Bricout JC (2010) Leveraging online social networks for people with disabilities in emergency communications and recovery. Int J Emergency Manage 7(1):59–74

Burgess J, Green J (2009) YouTube: online video and participatory culture. Polity Press, Oxford

Cha M, Kwak H, Rodriguez P, Ahn Y-Y, Moon S (2007) I Tube, You Tube, Everybody Tubes: analyzing the world's largest user generated content video system. In: Proceedings of the 7th ACM SIGCOMM conference on internet measurement. http://dl.acm.org/citation.cfm?id=1298309 (26.03.2017)

Chambers S (2003) Deliberative democracy theory. Ann Rev Polit Sci 6:307–326

Charlwood J (2012) Use of social media during flood events. In: Floodplain management association national conference papers. http://www.floodplainconference.com/papers2012.php (26.03.2017)

Cheng X, Dale C, Liu J (2008) Statistics and social network of YouTube videos. In: 2008 16th international workshop on quality of service. http://ieeexplore.ieee.org/document/4539688/?tp= &arnumber=4539688 (26.03.2017)

Czerniawska D (2012) Wykluczenie cyfrowe. Strukturalne uwarunkowania korzystania z internetu w Polsce i w województwie mazowieckim. Mazowiecka Jednostka Wdrażania Programów Unijnych, Warszawa

Delli Carpini MX, Lomax Cook F, Jacobs LR (2004) Public deliberation, discursive participation, and citizen engagement: a review of the empirical literature. Ann Rev Polit Sci 7:315–344

Działek J (2013) Perception of natural hazards and disasters. In: Bobrowsky PT (ed) Encyclopedia of natural hazards. Springer, Dordrecht, pp 756–759

Höppner C, Bründl M, Buchecker M (2010) Risk communication and natural hazards, CapHaz-Net WP5 Report, Swiss Federal Research Institute WSL. http://caphaznet.org/outcomes-results/ CapHaz-Net_WP5_RiskCommunication.pdf (25.02.2017)

Höppner C, Whittle R, Bründl M, Buchecker M (2012) Linking social capacities and risk communication in Europe: a gap between theory and practice? Nat Hazards 64(2):1753–1778

Houston JB, Hawthorne J, Perreault MF, Park EH, Goldstein Hode M, Halliwell MR, Turner McGowen SE, Davis R, Vaid S, McElderry JA, Griffith SA (2015) Social media and disasters: a functional framework for social media use in disaster planning, response, and research. Disasters 39(1):1–22

Hughes AL, St. Denis LA, Palen L, Anderson KM (2014) Online public communications by police & fire services during the 2012 Hurricane Sandy. In: Proceedings of the SIGCHI conference on human factors in computing systems, pp 1505–1514

Komac B, Ciglič R, Erhartič B, Gašperič P, Kozina J, Orožen Adamič M, Pavšek M, Pipan P, Volk M, Zorn M (2010) Risk education and natural hazards. CapHaz-Net WP6 report. Ljubljana. http://caphaz-net.org/outcomes-results/CapHaz-Net_WP6_Risk-Education (15.03.2015)

Lee KL, Meyer RJ, Bradlow ET (2009) Analyzing risk response dynamics on the web: the case of Hurricane Katrina. Risk Anal 29(12):1779–1792

Lindell MK, Perry RW (2004) Communicating environmental risk in multiethnic communities. Sage, Thousand Oaks, CA

Neubaum G, Rösner L, Rosenthal-von der Pütten AM, Krämer NC (2014) Psychosocial functions of social media usage in a disaster situation: a multi-methodological approach. Comput Hum Behav 34:28–38

Palen L, Hughes AL (2017) Social media in disaster communication. In: Rodríguez H, Donner W, Trainor J (eds) Handbook of disaster research. Handbooks of sociology and social research. Springer, Cham

Parkins JR, Mitchell RE (2005) Public participation as public debate: a deliberative turn in natural resource management. Soc Nat Resour 18:529–540

Roth F, Brönnimann G (2013) Focal report 8: risk analysis using the internet for public risk communication. Risk and Resilience Research Group, Center for Security Studies (CSS), ETH Zürich

Simonsen TM (2011) Categorising YouTube. Meide Kultur J Media Commun Res 51:72–93

Stanbrough L (2013) Internet, world wide web and natural hazards. In: Bobrowsky PT (ed) Encyclopedia of natural hazards. Springer, Dordrecht-New York, pp 563–565

Stanghellini PSL, Collentine D (2008) Stakeholder discourse and water management—implementation of the participatory model CATCH in a Northern Italian alpine subcatchment. Hydrol Earth Syst Sci 12:317–331

Steinführer A, Kuhlicke C, De Marchi B, Scolobig A, Tapsell S, Tunstall S (2009) Local communities at risk from flooding: social vulnerability, resilience and recommendations for flood risk management in Europe. FLOODsite, Leipzig

Tapsell S, McCarthy S, Faulkner H, Alexander M (2010) Social vulnerability and natural hazards. CapHaz-Net WP4 report. Flood Hazard Research Centre—FHRC, Middlesex University, London. http://caphaz-net.org/outcomes-results/CapHaz-Net_WP4_Social-Vulnerability2. pdf (15.05.2016)

Walls J, Pidgeon N, Weyman A, Horlick-Jones T (2004) Critical trust: understanding lay perceptions of health and safety risk regulation. Health Risk Soc 6(2):133–150

YouTube (2017) YouTube for press. https://www.youtube.com/yt/about/press/ (25.03.2017)

# Chapter 6
# Planning Documents as a Source of Information About Flood Risk

Floods cause serious damage to infrastructure and financial losses, which is largely due to inappropriate management of floodplains. To a large extent, this is associated with inadequate spatial planning in these areas, and very often even absence of any planning at all. The boundaries of flood zones have been delimited, and then included in planning documents, only for approximately 40% of flood-prone areas in Poland. In addition, the boundaries set within some small uncontrolled catchment areas (not included in permanent hydrological monitoring) often diverge from the actual reach of water recorded during catastrophic floods, with the flood zone usually much wider than that specified in the planning documents.

Incorporating reliable information about flood-prone areas in local planning documents should be treated as an element of communicating flood risk since it gives residents a means of learning which areas are dangerous and should be covered by development restrictions. According to information received from Security and Crisis Management Departments of the Regional Offices, about 86,500 residential buildings, 2600 public buildings and about 8750 km$^2$ of agricultural land in Poland are at risk of flooding (*Zagrożenia* … 2010).

## 6.1 Flood Hazard in Land-Use Planning Documents

In recent decades, European countries have modified their approach to flood risk management (Hooijer et al. 2004; Petrow et al. 2006; van Alphen and van Beek 2006). Some countries are increasingly moving away from traditional flood protection, e.g. Germany (DKKV 2004), the Netherlands (Vis et al. 2003; Roos and van der Geer 2008) and the United Kingdom (Tunstall et al. 2004). The construction of new technical infrastructure (including flood embankments, large water reservoirs) has been replaced by moving people away from the river and more rational spatial planning within new flood risk management strategies known as the *Making Space for Water* and *Making Room for the River*. They intend to stop new housing developments aimed at floodplains and leave more space, where floodwaters can spill freely. This approach is aimed at reducing flood damage, as there is no possibility of eliminating them completely (Pottier et al. 2005; van Heezik 2008; Krieger 2013). However, Wynn (2005), in her research conducted in the United Kingdom, noticed that the development pressure in urban areas hinders proper implementation of these strategies.

Rational and consistent spatial planning and flood risk management in areas at risk of flooding contribute to mitigating the effects of flooding (Howe and White 2004; White and Richards 2007). This is due to the limitation of intense development on flood-threatened areas, and thus limitation of the resulting damage (White and Richards 2007; Neuvel and van der Knaap 2010). This approach is in line with the Flood Directive adopted on 23 October 2007 by the European Parliament (2007/60/EC). Its purpose is to establish a framework for flood risk assessment in European countries, and then to manage it. The directive also required EU Member States to develop flood risk management plans and a preliminary flood risk assessment together with flood hazard and flood risk maps (de Moel et al. 2009).

In individual countries, the responsibility for risk management rests with the central government and in others with local authorities. For example, in Switzerland, the Federal Government provides financial, technical and scientific support (Ran and Nedovic-Budic 2016), while the cantons are responsible for identifying areas at risk of flooding and for risk management in these areas. However, they can transfer these tasks to municipalities. In turn, in England and Wales, it is the responsibility of the Environment Agency.

Flood hazard maps are nothing new as they already existed in some of European countries, as well as in the USA and Canada (van Alphen et al. 2008; de Moel and Aerts 2008). In the USA, the National Flood Insurance Program was introduced in 1968 (Burby 2001; de Moel et al. 2009), while in Europe, large flood prevention programmes began to operate from the 1990s (Høydal et al. 2000). The Flood Directive (2007/60/EC), however, introduced a requirement that all EU Member States should have them.

Previously, the obligation to develop a flood protection study in Poland was based on the provisions of the Act of 18 July 2001 on Water Law, pursuant to Article 79 para. 2. The director of the Regional Water Management Board (Regionalny Zarząd Gospodarki Wodnej, RZGW) was obliged to prepare an analysis determining the limits of floodwater coverage and the objectives of flood protection. For the upper Vistula river basin, the RZGW in Kraków began to develop such a study from 2003 (they were developed for individual catchments, e.g. Skawa, Soła and Raba Rivers). After the amendment of the Water Law Act and some other acts in 2011, the flood protection study was replaced with flood hazard and flood risk maps, and the areas of direct flood hazard designated in the previous study were named areas of special flood hazard (RZGW Kraków 2017).

The flood risk maps developed in individual EU Member States are used by different governments for different purposes. However, they are mainly used in emergency planning (e.g. evacuation) and in spatial planning. Their use in spatial planning differs from country to country (de Moel et al. 2009): flood risk maps in Sweden and Norway are an information tool for decision-makers, while in Finland and the United Kingdom there is a legal obligation to introduce zones threatened by floods in planning documents, even though development in these areas cannot be completely prohibited. In Poland, similarly to France and Germany, there is a ban on the development of buildings in areas of flood hazard. However, there are statutory options to exempt this prohibition. In Switzerland and Spain, local governments decide whether or not the flood zones set out on the maps are included in their spatial plans or not (Zimmerman et al. 2005; Cantos 2005).

The use of flood hazard and flood risk maps differs not only in legal terms but above all there are some differences in how these zones are designated, how information is presented, designed and disseminated. Although in the EU such an obligation stems from a single legal document, de Moel et al. (2009) have demonstrated that these maps vary widely between countries. Sometimes, they only present areas threatened by flooding; in other cases, this information is supplemented with the depth and speed of water, and in others, as in Ireland, we can see the flood zones from historical floods, and in Finland the extent of floods in the preceding decades that is between 1988 and 2006 (D'Haeseleer et al. 2006; de Moel et al. 2009). In addition, flood maps developed for the Czech Republic include a range of major historical floods (floods in 1997 and 2002), which have been significantly engraved on social memory. Similar maps were prepared for some regions in France and Finland (van Alphen and Passchier 2007). Such a solution could have been adopted in the Polish versions of flood maps (for the Odra basin showing 1997 flood, and for the Vistula basin—2010 flood).

Flood hazard maps and flood risk maps developed for Poland were provided by contractors to the local government administration unit, while individual persons were able to use them through the ISOK (Informatyczny System Ochrony Kraju = IT System of National Protection) portal for the first few years after they were made available. Maps for individual regions of the country were placed on sheets available for download in pdf format, but the time needed to reach individual sheets was relatively long (Głosińska 2014; Franczak and Listwan-Franczak 2017). Often, fragments of a given city or village were on several sheets prepared at a scale of 1:10,000, which made it difficult for the citizen, who did not necessarily know enough about local geography to use these sheets, to obtain information. In some countries (including UK, USA, France, the Netherlands), searching on maps is possible after entering the address with information about the flood risk, or lack thereof (Konieczny et al. 2017). In some countries (e.g. the Netherlands), additional information on how to behave before, during and after a disaster is provided. In the UK information about development requirements is added. These maps, due to their ease of access, are also used in national risk education campaigns in Ireland and the UK for the preparation of local communities in the event of floods (de Moel et al. 2009). From August 2018, flood risk maps in Poland have also been made available on geoportal.gov.pl. Facilitating access to these maps may in the future help citizens to raise awareness of the flood risk occurring in their neighbourhood.

Additionally, it is worth noting that maps showing areas at risk of flooding calculated according to a computer model, as suggested by de Moel et al. (2009), may be perceived by local communities as inaccurate and treated with suspicion. The extent of historical floods that the local population has experienced is much more readily perceived as accurate by the local population as information about them has been transmitted from generation to generation, even though as other studies in this book have shown, social memory about these floods and their flood zones is not perfect. One should also therefore consider introducing the actual areas of historical floods on flood risk maps as is done in the Czech Republic to strengthen risk communication via spatial planning documents. Noting the locations of important public institutions, including hospitals, as has been done in some regions in Germany, is another good practice in this field.

Similarly to other Western European countries, though with some delay and not to the full extent, there was also a change in the approach of Polish planners and decision-makers to flood protection. In order to reduce the adverse effects of floods, they started to shift from flood protection to flood risk mitigation (Konieczny et al. 2012; Merz et al. 2012; Bryndal 2014c). Previously, the focus had been on the development of hydrotechnical infrastructure with a view to ensuring flood safety. Huge sums of money had been spent to this end, but the losses associated with floods grow even bigger, as a result of an increase in the density of building development within flood-prone areas and the associated growth in the cumulative value of the property located there (Barrera et al. 2006; Biedroń and Bogańska-Warmuz 2012; Lorenc et al. 2012; Romanowicz et al. 2014). This is connected with the fact that technical protections often give the people living in their vicinity an illusory sense of security, which leads to investments within flood-prone areas, and once such protections are destroyed—to much higher losses (Dubicki and Gierczak 2011; Zaleski 2011; Konieczny et al. 2017).

The above experience has shown that full protection against flooding is impossible and that decisions taken should aim at minimising the economic losses caused by floods (Bryndal 2014c). In addition, the level of flood risk does not only depend on the flood hazard but also on the exposure and vulnerability of residents to floods (Kron 2012; Konieczny et al. 2012; Bryndal 2014c).

The problem of flood protection by means of the construction of hydrotechnical structures is even greater in small catchment areas, where such structures can only slightly reduce the height of the flood wave. Reservoirs built within such basins are located at points which 'close' catchment areas that are too small for the reservoirs to reduce significantly a flood wave as it builds up, and moreover they rapidly fill with sediments, reducing their retention capacity (Beuselinc et al. 2000; Bryndal 2014c).

Flood risk can be reduced most effectively through limitation of the flood exposure, which can be achieved by proper spatial planning. By following the principle of 'moving people away from flood-prone areas' (Żelaziński 2011), flood losses can be reduced significantly, and risk of flooding can be mitigated by preventing development within floodplains. The inundation of undeveloped land by floodwaters can (at most) alter the landform and temporarily change the landscape, causing no flood losses or damage (Franczak and Listwan-Franczak 2015). When flood-prone areas are nevertheless developed, efforts should be made to reduce the vulnerability of populations living there. This can be done through flood education addressed to residents, creating flood warning systems and protecting buildings threatened by floods (Monz and Grunfest 2002; Konieczny et al. 2012, 2016; Merz et al. 2012; Działek and Biernacki 2014; Bryndal 2014c).

## 6.2 Legal Aspects of Floodplain Management in Poland

Reducing exposure to floods by means of appropriate planning is one of the most effective ways of decreasing flood losses and devastation (Grocki and Eliasiewicz 2001; Wołoszyn 2006; Żelaziński 2007; Wheater and Evans 2009; Kaźmierczak and Cavan 2011; Richert et al. 2011; Ristic et al. 2012). An important step in support of such activities was the adoption at EU level of the *Flood Directive* (Directive of the European Parliament and of the Council of 23 October 2007 on the assessment and management of flood risks), which sets forth the rules on flood risk assessment and management. The Directive required each Member State to prepare a preliminary flood risk assessment (*Raport … 2011*), and then, on its basis, to prepare flood hazard and flood risk maps (Głosińska 2014). Both types of maps were to be prepared for areas found to pose high flood risk or where such risk is likely (Kurczyński 2012; *Mapy … 2015*). The maps included three scenarios based on the extent of floodwaters involving the likelihood of the occurrence of the so-called decadal water ($Q_{10\%}$), centennial water ($Q_{1\%}$) and semi-millennial water ($Q_{0.2\%}$) (*Mapy … 2015*).

Furthermore, in areas with flood infrastructures, areas were identified which were at risk of flooding in the event of damage to flood embankments or dams (Article 88d, *Water Law* 2001). The boundaries of these areas were delimited by mathematical hydraulic modelling, based on data from a numerical terrain model with a height accuracy of 10–15 cm. In addition, the flood hazard maps showed the flood depth, and in large cities, information about the direction and velocity of the water flow (*Mapy … 2015*). However, the maps have certain drawbacks. For example, they assume that areas protected by flood embankments would accommodate the flood wave and that water would not inundate the landside of the levee. As a result, the flood hazard area modelled within downstream non-embanked reaches of big rivers is wider than it would occur in cases when flood embankments were actually breached or overtopped.

In Polish planning documents, the flood hazard zone is assumed to correspond to what is referred to as 'centennial water' ($Q_{1\%}$) (Łyp 2005; Biedroń and Walczykiewicz 2006; Ryłko 2006; Lachowska 2016; Franczak et al. 2016). Most probably, this is because of the fact that areas beyond this range are not very likely to be affected by floods. The legislator did not specify the criterion for delimiting areas exposed to direct and potential flood hazard, and therefore a range of other criteria were also used (Lachowska 2016). Very often, the zones were delimited on the basis of existing land development, with the area at risk of flooding mapped accordingly (Kitowski 2010; Ryłko 2010). However, for some towns which had suffered extensive damage and high losses during previous floods, the flood hazard area was expanded in comparison to earlier planning documents. For example in Bogatynia, the municipal urban development study adopted in 2014 delimited the flood hazard area for the Miedzianka River in correspondence to the range of the floodwaters during the catastrophic flood in 2010 (*Studium ... 2014*). The area covers a very large part of the city and is much wider than the one previously set—following the revision, it corresponds to the extent of the most catastrophic floods which would cause huge losses in the Miedzianka River valley in the past (Franczak and Listwan-Franczak 2016a). In Tuchów, the flood range in the 2014 local urban development study was delimited by reference to the so-called 'centennial water', but the increased flood hazard area was determined according to the range of the 2010 flood (*Miejscowy ... 2014*).

The options for delimiting the area were relatively wide ranging as it was mapped on the basis of Flood Protection Studies prepared by competent Regional Water Management Authorities. The latter documents distinguish between more than one types of zone which comprise areas exposed to various direct or indirect flood hazards. In addition to the boundaries of floodwater with the probability $Q_{1\%}$, they define zones having the following ranges: $Q_{50\%}$, $Q_{20\%}$, $Q_{10\%}$, $Q_{5\%}$, $Q_{3.33\%}$ and $Q_{0.2\%}$ (Biedroń and Walczykiewicz 2006; Lachowska 2016). In addition, for embanked areas, the studies also mapped the boundaries of zones exposed to the risk of 'centennial' ($Q_{1\%}$) and 'semi-millennial' water ($Q_{0.2\%}$), which could be affected in the event flood banks were damaged or completely destroyed.

In 2011, the Water Law was revised. Pursuant to the amended legislation, direct flood hazard areas (obszary bezpośredniego zagrożenia powodzią) were defined as those corresponding to increased flood hazard areas (obszary szczególnego zagrożenia powodzą) (Article 17(1) of the *Water Law* 2001). In light of Article 9(1) Point 6c of the Law, increased flood hazard areas are defined as land where the likelihood of the occurrence of flooding is moderate and corresponds to 'centennial water' ($Q_{1\%}$). In addition, the subsequent points of the above Article 9(1) of the Law provide that such areas should also comprise land where the probability of flooding is high and is once per 10 years ($Q_{10\%}$), as well as inter-embankment zones, and the technical access zone. Thus, the criterion for determining zones at risk of flooding was standardised, but it must be remembered that urban development studies prepared before the revision of the Water Law still include zones whose boundaries were determined according to varying criteria.

The situation in mountain areas is slightly different. As is shown by studies of many authors (German 1998; Izmaiłow et al. 2008; Bryndal 2008, 2011, 2014b, 2015a, b, c; Gorczyca and Krzemień 2008; Gorczyca and Wrońska-Wałach 2008; Bryndal et al. 2010a, b, c; Franczak and Listwan 2015; Franczak and Listwan-Franczak 2016a, b; Witkowski and Franczak 2016), many relatively small mountain and sub-montane basins experience floods having a much greater extent than floods likely to involve 'centennial water' ($Q_{1\%}$). During the so-called 'Millennial Flood' in 1997, the discharges of some streams in the Kłodzko Valley involved likelihoods ranging from $Q_{0.5\%}$ ('bi-centennial' water) to $Q_{0.002\%}$ (once every 50,000 years) (Dubicki et al. 1999a, b). Also, studies in the Carpathians (Bryndal 2014a) have shown that flash floods are most common in small catchments (up to 100 km$^2$), with the maximum discharges often much higher than the 'centennial water' ($Q_{1\%}$). As a result, especially small and medium mountain catchment areas (up to approx. 150 km$^2$) see floods which diverge highly from the range of flooding observed during catastrophic floods (Bryndal 2014c, 2015c). In the catchment areas researched by Franczak et al. (2016), during various floods, the width of the floodplain was 1.5–2 times wider than that delimited in the respective planning documents, and in extreme cases—almost four times wider. In small mountain basins, as a result of flash floods nearly the entire valley bottoms are flooded. In such catchment areas, on account of the stream channels cutting deeply into the terrain, the determined area likely to be flooded with centennial or greater water often comprised the channel itself, or the channel along with a narrow strip of the floodplain. However, once they occurred there, flash floods had a much lower probability of the occurrence and inundated much wider areas.

For this reason, as is proposed by Bryndal (2014c, 2015b), within small catchment areas, the flood zone should be delimited in correspondence to the so-called 'maximum credible water level rise' (Ozga-Zielińska et al. 2003) or be based on what is referred to as 'envelope equations' for maximum discharges (Bartnik and Jokiel 2012; Bryndal 2014d). The above methods can be used for determining the probability of the maximum discharge, and thus, establishing the maximum zone likely to be inundated by floodwaters.

Delimiting such zones could also contribute to reducing flood vulnerability by making residents of flood-prone areas aware that the land where they live is likely to be affected by floods (Franczak and Listwan 2015; Franczak et al. 2016). When making and approving investment decisions, local governments, which are aware of the extent of potential floodwaters, should plan the construction of technical and municipal infrastructure in a more rational way, and they should prepare flood response and recovery plans (Franczak and Listwan 2015; Franczak et al. 2016).

# References

Barrera A, Llasat MC, Barriendos M (2006) Estimation of extreme flash flood evolution in Barcelona County from 1351 to 2005. Nat Hazard Earth Syst Sci 6:505–518

Bartnik A, Jokiel P (2012) Geografia wezbrań i powodzi rzecznych. Wyd. Uniwersytetu Łódzkiego, Łódź

Beuselinc L, Steegen A, Govers G, Nachtergaele J, Takken I, Poesen J (2000) Characteristics of sediment deposits formed by intense rainfall events in small catchments in the Belgian Loam Belt. Geomorphology 32:60–82

Biedroń I, Bogańska-Warmuz R (2012) Powódź 2010—analiza strat i szkód powodziowych w Polsce. Gospodarka Wodna 4:147

Biedroń I, Walczykiewicz T (2006) Mapy zagrożenia powodziowego w kontekście jego oceny i planowania przestrzennego. Wiadomości Instytutu Meteorologii Gospodarki Wodnej 29(3–4):59–69

Bryndal T (2008) Parametry zlewni, w których wystąpiły lokalne powodzie. Ann UMCS Sect B Geogr Geol Mineral Petrogtaph 63:177–200

Bryndal T (2011) Identyfikacja małych zlewni podatnych na formowanie gwałtownych wezbrań na przykładzie Pogórza Dynowskiego, Strzyżowskiego i Przemyskiego. Prz Geogr 83(1):5–26

Bryndal T (2014a) Identyfikacja małych zlewni podatnych na formowanie gwałtownych wezbrań w Karpatach Polskich. Prace Monograficzne UP, 690. Wyd. Naukowe Uniwersytetu Pedagogicznego, Kraków

Bryndal T (2014b) Parametry hydrologiczne gwałtownych wezbrań opadowo-nawalnych w małych zlewniach, w polskiej, słowackiej i rumuńskiej części Karpat. Prz Geogr 86(1):5–21

Bryndal T (2014c) Powodzie błyskawiczne w małych zlewniach karpackich—wybrane aspekty zarządzania ryzykiem powodziowym. Ann Univ Paedagog Crac Stud Geogr 7(170):69–80

Bryndal T (2014d) Znaczenie map zagrożenia oraz ryzyka powodziowego w ograniczeniu skutków powodzi błyskawicznych w miastach. In: Ciupa T, Suligowski R (eds) Woda w mieście. IG UJK w Kielcach, Kielce, pp 29–37

Bryndal T (2015a) Local flash floods in Central Europe: a case study of Poland. Norsk Geogr Tidsskrift—Norw J Geogr 69(5):288–298

Bryndal T (2015b) Obszary predysponowane do wystąpienia gwałtownych wezbrań w Karpatach w kontekście przeciwdziałania ekonomicznym skutkom powodzi błyskawicznych. Ann Univ Paedagog Crac Stud Geogr 9:24–37

Bryndal T (2015c) The river systems in small catchments in the context of the Horton's and Schumm's laws—implication for hydrological modelling: the case study of the Polish Carpathians. Quaest Geogr 34(1):85–98

Bryndal T, Cabaj W, Suligowski R (2010a) Gwałtowne wezbrania potoków Kisielina i Niedźwiedź w czerwcu 2009 r. (Pogórze Wiśnickie). In: Barwiński M (ed) Obszary metropolitarne we współczesnym środowisku geograficznym. Wydział Nauk Geograficznych Uniwersytetu Łódzkiego, Łódź, pp 337–348

Bryndal T, Cabaj W, Suligowski R (2010b) Hydrometeorologiczna interpretacja gwałtownych wezbrań małych cieków w źródłowej części Wielopolki w dniu 25 czerwca 2009 roku. Monogr Komitetu Inżynierii Środowiska PAN 69:81–91

Bryndal T Cabaj W, Gębica P, Kroczak R (2010c) Gwałtowne wezbrania spowodowane nawalnymi opadami deszczu w zlewni potoku Wątok (Pogórze Ciężkowickie). In: Ciupa T, Suligowski R (eds) Woda w badaniach geograficznych. Instytut Geografii Uniwersytetu Jana Kochanowskiego, Kielce, pp 307–319

Burby RJ (2001) Flood insurance and floodplain management: the US experience. Glob Environ Change Part B Environ Hazards 3:111–122

Cantos JO (2005) Country report—Spain. In: Greiving S, Fleischhauer M, Wanczura S (eds) Report on the European scenario of technological and scientific standards reached in spatial planning versus natural risk management. ARMONIA Project, Dortmund

D'Haeseleer E, Vanneuville W, Van Eerdenbruch K, Mostaert F (2006) Gebruik van overstromingskaarten voor verschilende watergerelateerde beheers -en beleidsinstrumenten. Water, September–Oktober: 1–5

de Moel H, Aerts JCJH (2008) Flood maps in Europe: a comparative evaluation of methods, availability and application. In: Proceedings of the 4th international symposium on flood defence, vol 28, Toronto, Canada, 6 May 2008, pp 1–11

de Moel H, van Alphen J, Aerts JCJH (2009) Flood maps in Europe—methods, availability and use. Nat Hazards Earth Syst Sci 9:289–301

DKKV (2004) Flood risk reduction in Germany—lessons learned from the 2002 disaster in the Elbe region. Deutsches Komitee f\`ur Katastrophenvorsorge e.V. (DKKV), Bonn, p 29e

Dubicki A, Gierczak J (2011) Analiza skuteczności wałów przeciwpowodziowych. In: Dubicki A, Słota H, Zieliński J (eds) Dorzecze Odry: monografia powodzi lipiec 1997. IMGW, Warszawa, pp 143–158

Dubicki A, Borowicz A, Turzańska-Chrobak B, Gierczak J, Lisowski J (1999a) Rzeczowe i finansowe straty powodziowe. In: Dubicki A, Słota H, Zieliński J (eds) Dorzecze Odry: monografia powodzi lipiec 1997. IMGW, Warszawa, pp 175–190

Dubicki A, Słota H, Zieliński J (eds) (1999b) Dorzecze Odry: monografia powodzi lipiec 1997. IMGW, Warszawa

Działek J, Biernacki W (2014) Wrażliwość społeczna na klęski żywiołowe—ujęcia teoretyczne i praktyka badawcza. Prace Stud Geogr 55:25–39

Franczak P, Listwan K (2015) Ryzyko powodziowe w małych zlewniach górskich a sposoby zagospodarowania obszarów zalewowych zapisane w aktach planistycznych: studium przypadku Makowa Podhalańskiego i Kasinki Małej. In: Liro J, Liro M, Krąż P (eds) Współczesne problemy i kierunki badawcze w geografii, vol 3. IGiGP UJ, Kraków, pp 45–61

Franczak P, Listwan-Franczak K (2015) Zmiany geomorfologiczne i krajobrazowe zachodzące w małych zlewniach górskich pod wpływem katastrofalnych wezbrań i ich trwałość w krajobrazie. Probl Ekologii Krajobrazu 39:33–44

Franczak P, Listwan-Franczak K (2016a) Powódź w zlewni Miedzianki (zlewnia Nysy Łużyckiej) w sierpniu 2010 roku. Dobra praktyka w redukcji ryzyka powodziowego w małych zlewniach górskich, w których wystąpiła powódź błyskawiczna. In: Franczak P, Krąż P, Liro J, Liro M, Listwan-Franczak K (eds) Współczesne problemy i kierunki badawcze w geografii, vol 4. IGiGP UJ, Kraków, pp 55–84

Franczak P, Listwan-Franczak K (2016b) Występowanie powodzi błyskawicznych w miastach położonych na przedpolu gór na przykładzie Bogatyni (Sudety). In: Hejduk L, Kaznowska E (eds) Hydrologia zlewni zurbanizowanych. Monografie Komitetu Gospodarki Wodnej Polskiej Akademii Nauk, KGW PAN. IMGW-PIB, Warszawa, pp 125–137

Franczak P, Listwan-Franczak K (2017) Akty planistyczne jako źródło informacji o ryzyku powodziowym. In: Działek J, Biernacki J, Konieczny R, Fiedeń Ł, Franczak P, Grzeszna K, Listwan-Franczak K (eds) Zanim nadejdzie powódź. Wpływ wyobrażeń przestrzennych, wrażliwości społecznej na klęski żywiołowe oraz komunikowania ryzyka na przygotowanie społeczności lokalnych do powodzi. IGiGP UJ, Kraków, pp 373–389

Franczak P, Listwan-Franczak K, Działek J, Biernacki W (2016) Planowanie przestrzenne na obszarach zalewowych w zlewniach górskich różnego rzędu w dorzeczu górnej Wisły oraz górnej i środkowej Odry. Prace Stud Geogr 61(4):25–45

German K (1998) Przebieg wezbrania powodzi 9 lipca 1997 roku w okolicach Żegociny oraz ich skutki krajobrazowe. In: Grela J, Starkel L (eds) Powódź w dorzeczu górnej Wisły w lipcu 1997 roku. Materiały z konferencji naukowej w Krakowie 7–9 maja 1998. Kraków, pp 177–183

Głosińska E (2014) Spatial planning in floodplains for implementation by the floods directive in Poland. Geogr Pol 87(1):127–142

Gorczyca E, Krzemień K (2008) Morfologiczne skutki ekstremalnego zdarzenia opadowego w Tatrach Reglowych w czerwcu 2007 r. Land Anal 8:21–24

Gorczyca E, Wrońska-Wałach D (2008) Transformacja małych zlewni górskich podczas opadowych zdarzeń ekstremalnych (Bieszczady). Land Anal 8:25–28

Grocki R, Eliasiewicz R (2001) Zagospodarowanie terenów zalewowych. Biuro Koordynacji Projektu Banku Światowego, SAFEGE, Wrocław

Hooijer A, Klijn F, Pedroli GBM, Van Os AG (2004) Towards sustainable flood risk management in the Rhine and Meuse river basins: synopsis of the findings of IRMA-SPONGE. River Res Appl 20:343–357

Howe J, White I (2004) Like a fish out of water: the relationship between planning and flood risk management in the UK. Plan Pract Res 19(4):415–425

Høydal ØA, Berg H, Haddeland I, Petterson LE, Voksø A, Øydvin E (2000) Procedures and guidelines for flood inundation maps in Norway. Potsdam, Germany, pp 252–261

Izmaiłow B, Kamykowska M, Krzemień K (2008) Geomorfologiczna rola katastrofalnych wezbrań w transformacji górskiego systemu korytowego na przykładzie Wilszni (Beskid Nicki). In: Izmaiłow B (ed) Przyroda—człowiek—Bóg. IGiGP UJ, Kraków, pp 69–81

Kaźmierczak A, Caven G (2011) Surface water flooding risk to urban communities: analysis of vulnerability, hazard and exposure. Landscape Urban Plan 103(2):185–197

Kitowski K (2010) Dyrektywa powodziowa a prewencyjne planowanie przestrzenne. Prz Komunalny 7:48–51

Konieczny R, Siudak M, Bogdańska-Warmuz M, Madej P, Walczykiewicz T (2012) Opracowanie systemu zapobiegania i sposoby ograniczania skutków powodzi oraz zasad funkcjonowania systemu ostrzeżeń. In: Lorenc H (ed) Wpływ zmian klimatu na środowisko, gospodarkę i społeczeństwo, t. 3: Klęski żywiołowe a bezpieczeństwo kraju. IMGW-BIP, Warszawa, pp 281–303

Konieczny R, Działek J, Siudak M, Biernacki W (2016) Działania właścicieli domów dla ograniczenia skutków powodzi oraz ich motywacje. In: Walczykiewicz T (ed) Problemy planowania w gospodarce wodnej i oceny stanu hydromorfologicznego rzek. IMGW-PIB, Warszawa, pp 171–189

Konieczny R, Kundzewicz ZW, Siudak M, Działek J, Biernacki W (2017) Education and information as a basis for flood risk management—practical issues. In: Tyszka T, Zielonka P (eds) Large risk with low probabilities: perceptions and willingness to take preventive measures against flooding. IWA Publishing, London, pp 177–201

Krieger K (2013) The limits and variety of risk-based governance: the case of flood management in Germany and England. Regul Gov 7:236–257

Kron W (2012) Changing flood risk a reinsurance's viewpoint. In: Kundzewicz ZW (ed) Changes in flood risk in Europe, vol 10. SCS Press, IAHS Special Publication, Wallingford, Oxfordshire, pp 459–477

Kurczyński Z (2012) Mapy zagrożenia powodziowego i mapy ryzyka powodziowego a Dyrektywa Powodziowa. Arch Fotogrametrii Kartogr Teledetekcji 23:209–217

Lachowska E (2016) Zmiany zagospodarowania obszarów zalewowych i ich wpływ na poziom ryzyka powodziowego w miastach nadodrzańskich w Polsce. Bogucki Wydawnictwo Naukowe, Poznań

Lorenc H, Cebulak E, Głowacki B, Kowalewski M (2012) Struktura występowania intensywnych opadów deszczu powodujących zagrożenie dla społeczeństwa, środowiska i gospodarki Polski. In: Lorenc H (ed) Wpływ zmian klimatu na środowisko, gospodarkę i społeczeństwo, t. 3: Klęski żywiołowe a bezpieczeństwo kraju. IMGW–BIP, Warszawa, pp 7–32

Łyp P (2005) Problematyka wodna w planowaniu przestrzennym miast. Poradnik dla urbanistów. Centralny Ośrodek Informacji Budownictwa, Warszawa

Mapy zagrożenia powodziowego i mapy ryzyka powodziowego (2015) http://www.isok.gov.pl/pl/mapy–zagrozenia–powodziowego–i–mapy–ryzyka–powodziowego (5.12.2015)

Merz B, Kundzewicz ZW, Delgado J, Hundecha Y, Kreibich H (2012) Detection and attribution of changes in flood hazard and risk. In: Kundzewicz ZW (ed) Changes in flood risk in Europe, vol 10. SCS Press, IAHS Special Publication, Wallingford, Oxfordshire, pp 435–458

Miejscowy plan zagospodarowania przestrzennego Gminy Tuchów, dla miasta Tuchów—II etap zmiany z 29 stycznia 2014 r. (2014) http://sip.tuchow.pl/konfiguracja/dok/mpzp_20140129.pdf (5.12.2015)

Monz BE, Grunfest E (2002) Flash flood mitigation: recommendations for research and applications. Nat Hazards 4:15–22

Neuvel JMM, van der Knaap W (2010) A spatial planning perspective for measures concerning flood risk management. Int J Water Resour Dev 26(2):283–296

Ozga-Zielińska M, Kupczyk E, Ozga-Zieliński B, Suligowski R, Niedbała J, Brzeziński J (2003) Powodziogenność rzek pod kątem bezpieczeństwa budowli hydrotechnicznych i zagrożenia powodziowego. Materiały Badawcze IMGW. Hydrol Oceanol 29

Petrow T, Thieken AH, Kreibich H, Bahlburg CH, Merz B (2006) Improvements on flood alleviation in Germany: lessons learned from the Elbe flood in August 2002. Environ Manage 38:717–732

Pottier N, Penning-Rowsell E, Tunstall S, Hubert G (2005) Land use and flood protection: contrasting approaches and outcomes in France and in England and Wales. Appl Geogr 25(1):1–27

Ran J, Nedovic-Budic Z (2016) Integrating spatial planning and flood risk management: a new conceptual framework for the spatially integrated policy infrastructure. Comput Environ Urban Syst 57(2016):68–79

Raport z wykonania wstępnej oceny ryzyka powodziowego (2011) http://www.kzgw.gov.pl/pl/wstepna–ocena–ryzyka–powodziowego.html (2.12.2015)

Richert E, Bianchin S, Hellmeier H, Merta M, Seidler C (2011) A method for linking results from an evaluation of land use scenarios from the viewpoint of flood prevention and nature conservation. Landscape Urban Plan 103(2):118–128

Ristic R, Kostadinov S, Abolmasov B, Dragicevic S, Trivan G, Radic B, Trifunovic M, Radosavljevic Z (2012) Torrential floods and town and country planning in Serbia. Nat Hazards Earth Syst Sci 12(1):23–35

Romanowicz RJ, Nachlik E, Januchta-Szostak A, Starkel L, Kundzewicz ZW, Byczkowski A, Kowalczak P, Żelaziński J, Radczuk L, Kowalik P, Szamałek K (2014) Zagrożenia związane z nadmiarem wody. Nauka 1:123–148

Roos A, Van der Geer I (2008) New approaches for flood risk management in the Netherlands. In: Proceedings of the 4th international symposium on flood defence, Toronto, Canada, 6–8 May 2008

Ryłko A (2006) Zagrożenie powodzią w planowaniu przestrzennym, Przestrzeń. Magazyn planowania przestrzennego, 19, 5

Ryłko A (2010) Proces uzgadniania dokumentów planistycznych z zakresu zagospodarowania przestrzennego—gromadzenie informacji o obszarach bezpośrednio zagrożonych powodzią. In: Więzik K (ed) Prawne, administracyjne i środowiskowe uwarunkowania zagospodarowania dolin rzecznych. Wyd. Wyższej Szkoły Administracji, Bielsko-Biała, pp 177–190

RZGW Kraków (2017) http://www.krakow.rzgw.gov.pl/wodypolskie_old/index.php?option=com_content&view=article&id=1040&Itemid=245&lang=pl (10.08.2017)

*Studium uwarunkowań i kierunków zagospodarowania przestrzennego miasta i gminy Bogatynia z 30 października 2014 r*. (2014) http://bip.bogatynia.pl/?a=6728 (27.12.2015)

Tunstall SM, Johnson CL, Pennning Rowsell EC (2004) Flood hazard management in England and Wales: from land, drainage to flood risk management. In: Proceedings of the world congress on natural disaster mitigation, New Delhi, India, 19 Feb 2004, pp 447–454

van Alphen J, Passchier R (2007) Atlas of flood maps, examples from 19 European countries, USA and Japan. Ministry of Transport, Public Works and Water Management, The Hague, Netherlands. Prepared for EXCIMAP, available at: http://ec.europa.eu/environment/water/floodrisk/floodatlas/index.htm

van Alphen J, van Beek E (2006) From flood defence to flood management—prerequisites for sustainable flood management. In: van Alphen J, van Beek E, Taal M (eds) Floods, from defence to management. Taylor & Francis Group, London

van Alphen J, Martini F, Loat R, Slomp R, Passchier R (2008) Flood risk mapping in Europe, experiences and best practices. In: Proceedings of the 4th international symposium on flood defence, vol 150, Toronto, Canada, 6 May 2008, pp 1–8

van Heezik A (2008) Battle over the rivers: two hundred years of river policy in the Netherlands. Van Heezik Beleidsresearch in association with the Dutch Ministry of Transport, Public Works and Water Management, The Hague

Vis M, Klijn F, De Bruijn KM, van Buuren M (2003) Resilience strategies for flood risk management in the Netherlands. Int J River Basin Manag 1:33–39

Wheater H, Evans E (2009) Land use, water management and future flood risk. Land Use Policy 26(10):S251–S262

White I, Richards J (2007) Planning policy and flood risk: the translation of national guidance into local policy. Plan Pract Res 22(4):513–534

Witkowski K, Franczak P (2016) Gwałtowne wezbranie spowodowane nawalnymi opadami deszczu w zlewni potoku Zygodówka (Pogórze Wielickie) w 2014 roku. In: Franczak P, Krąż P, Liro J, Liro M, Listwan-Franczak K (eds) Współczesne problemy i kierunki badawcze w geografii, vol 4. IGiGP UJ, Kraków, pp 309–327

Wołoszyn E (2006) Oddziaływanie powodzi na środowisko. In: Bednarczyk S, Jarzębińska T, Mackiewicz S, Wołoszyn E (eds) Vademecum ochrony przeciwpowodziowej. KZGW, Gdańsk

Wynn P (2005) Development control and flood risk: analysis of local planning authority and developer approaches to PPG25. Plan Pract Res 20(3):241–261

*Zagrożenia okresowo występujące w Polsce* (2010) Rządowe Centrum Bezpieczeństwa, Warszawa. http://rcb.gov.pl/wp–content/uploads/2011/02/zagr_okres1.pdf (12.12.2016)

Zaleski J (2011) Odra w kontekście sytuacji zagrożenia powodziowego i awarii budowlanych. In: XXV Konferencja Naukowo-Techniczna "Awarie Budowlane 2011", Międzyzdroje, 24–27 maja 2011, pp 321–334. http://www.awarie.zut.edu.pl/files/ab2011/referaty/T1_02_Powodzie_ w_Polsce_zniszczenia_i_profilaktyka/07_Zaleski_J_Odra_w_kontekscie_sytuacji_zagrozenia_ powodziowego_i_awarii_budowlanych.pdf (27.12.2015)

Żelaziński J (2007) Rola map terenów zalewowych w planowaniu ochrony przeciwpowodziowej. In: Nieznański P (ed) Bezpieczna granica nad Odrą. WWF Polska, Wrocław

Żelaziński J (2011) Nauczmy się żyć z powodziami. Let's learn to live with flooding. Infos, Biuro Analiz Sejmowych 2:1–4

Zimmerman M, Pozzi A, Stoessel F (2005) Vademecum—hazard maps and related instruments, the Swiss system and its application abroad. PLANAT, Bern, Switzerland. Available at: http://162. 23.39.120/dezaweb/ressources/resource_en_25123

# Chapter 7
# Discussion and Conclusions

When summarising findings on social memories of flood spatial image, some important outcomes should be discussed. The relationship between the land relief and the shape of the river valley and the morphology of the town's building development seems to come to the fore. Each of the towns has its own characteristic scenic spots, in each of them the flood caused different devastations, and in each of them it constrained the residents' activity. The most important determinants of the differences in the perceived size and extent of the flood by individual residents are as follows:

1. The nature of the phenomenon—rapid flood or less abrupt flooding—and in particular long-lasting difficulties caused by floodwaters, make residents better remember the flood. In Wojcieszów, everything happened so fast that nobody was able to look at the actual size of the phenomenon, not to mention additional constraints on their moving around the flooded area.
2. The topographic characteristics of the town's area and the resultant ease of floodwater observation. The latter was the easiest in Bieruń and then in Czarna, while watching the flood in Wojcieszów was practically impossible, precisely because of the location of the road, the only transport route connecting the lengthwise extended area of the town.
3. Age and associated communication-related and cognitive limitations of the flood victim. The smallest areas were marked by persons aged 70 and more who had limited possibilities of moving about the area on account of their old age and lower mobility. Their image of the flood was literally confined to the view from their own window. It is not expanded later on due to limited contact with their neighbours or other residents.
4. The 'proximity' of the problem posed by the flood—the most accurately delimited flood areas were those marked by members of the volunteer fire service, as well as the town residents and other persons who were directly engaged in the recovery, for example, employees of construction companies.

Across the localities, the flood lasted from the time the towns were inundated to the moment when the routine of everyday life was restored (rather than when the

J. Działek et al., *Understanding Flood Preparedness*, SpringerBriefs in Geography, https://doi.org/10.1007/978-3-030-04594-4_7

water subsided), so this differed from one part of the town to another. The notions or concepts retained in residents' memory often include the most characteristic buildings, structures or events, which they tend to refer to when remembering the flood in an effort to reconstruct its course. Reoccurrence of heavy rain or raised water levels in the river bring back flood-related memories for nearly all the respondents.

The most significant change observed for all the study areas after the 2 years between the first and second round of the surveys is the shrinkage of the area indicated by the respondents. It is difficult to establish the definitive reasons behind the mechanism, but clearly, the lapse of time contributed to the reduction in the size of the area perceived as previously flooded. In addition to the memory of the spatial extent of the flood, the familiarity with the area in general, one's local activity, and consequently, one's ability to take a broader look at the surroundings are significant. The more a given resident perceives space as a place (Tuan 2002), the better they remember a flood, the stronger their memories, and the greater the precision and accuracy of their indications of the flood area.

One way to strengthen the spatial dimension of flood memories would be the inclusion of flood maps with designated hazard zones in land-use planning documents. Flood hazard and risk maps calculated according to Flood Directive procedures, though often difficult to understand for non-experts, are becoming more accessible to citizens. However, they are not translated in a clear manner in official documents prepared in each municipality, resulting in weaker risk awareness and a higher vulnerability to floods. Moreover, in some cases there are significant discrepancies between the floodplains designated in land-use plans and the actual size of recent historic floods. Therefore, we propose that the extent of these devastating natural disasters should be included in spatial planning documents, fulfilling the role of a flood risk communication tool accessible at local scale.

As a rule, experiencing a flood makes people undertake flood preparation actions. Households differ considerably in terms of the scale of such activities: from households which take no actions at all and feel unprepared for another flood, to households which take a range of protective measures and feel that their protections are sufficient.

It turns out that the above differences can be best explained by the resources of human and social capital available to individual households. Those whose members were better educated, more often engaged in the activity of local associations, and were more closely linked with the local community proved to be better prepared for a potential disaster. This confirms the importance of knowledge about dangerous phenomena and ways to prepare for them, which can be passed on through local social networks. It also highlights the significance of risk communication, which can complement the aforementioned knowledge, and of relying on lay knowledge of local residents. The Internet and social media can play an important role in this regard.

The study has also revealed the crucial importance of motivational factors, such as belief in the possibility of reducing flood damage. Thus, risk communication should involve showing that flood loss mitigation is actually possible (at least to some extent) and should include presentation of various flood protection solutions.

The analysis has also shown that elderly persons are once more a group requiring special attention and support on account of their increased social vulnerability. Another category which should be dedicated special attention is tenants. At the same time, families with children seem to be more motivated to undertake preparatory actions; however, it does not mean there are no vulnerable households among them, or that they are not more vulnerable during a disaster or after it.

We believe that vulnerability assessment should not only identify potentially vulnerable groups or individuals likely to find themselves in a 'vulnerable situation' but also indicate ways to communicate with these groups to improve their knowledge, social networks and motivation.

However, two-way communication between people at risk and agencies responsible for flood risk management is clearly insufficient. Often, materials available online prepared by institutions located 'farther' from the actual flood, such as regional offices, research and scientific centres, or government agencies are written in inaccessible language, have no clear addressees, and since they often miss a local context, they prove to have low informative value for an average reader.

The activities of the science and research sector, which are focused on short-term, thematically diversified projects, often with no actual audience or space for application, play a minor role in the distribution of information, good practices and solutions adopted in other parts of the world. A promising sign is the observable cooperation between experts from the research sector and institutions specialising in floods (KZGW, IMGW), which results in the production of educational and guidance materials. However, the materials are still poorly disseminated among employees of municipalities and residents of at-risk zones.

The Internet is used by municipalities to keep residents informed about key developments when the hazard is still present or in the wake of a flood (e.g. dates when waste will be removed, points where requests for compensation from the state can be filed) and to educate residents (leaflets on what steps to follow in the face of a crisis, the conditions on the use of text message notification systems, as well as—unfortunately quite rarely—publication of flood hazard maps). The way in which contents published on municipalities' websites are presented also reveals a lack of regularly collected and systematised knowledge about past phenomena, which could be published—also on their websites—and be an exhaustive archive of events related to the occurrence of floods in a given locality among members of the local community. It is clear from interviews with local opinion leaders that it would be much easier for schools, local community centres, associations, etc. to create flood resilience culture in heavily threatened areas if these organisations had access to such contents and materials that would enable them to strengthen social memory of floods.

For residents of flooded areas, the Internet has become a space where they can share their experiences (especially since 2000s), notably via thematic forums (which unfortunately also disappear very quickly) and such social networks as YouTube. They use Internet to exchange information both immediately during or shortly after a disaster, and sometimes long after it is gone. Naturally, information posted there—mainly photographs and videos—are subjective, selective and tend to document events in a spontaneous way. The very publication and viewing of materials showing floods from the perspective of Internet users (participants) is a very important and valuable proof of the impact of the phenomena on the public, but such information is not systematised. Additionally, their use in flood risk communication and education is practically non-existent. Public institutions create their own materials without referring to the knowledge generated by people. It proves how far we are from undertaking a proper dialogue in flood risk management that would open more possibilities for two-way communication.

The information presented in this book describes the current risk information- and education-related situation faced by residents of localities affected by floods. The resultant picture is that of a highly dispersed system of contents available on websites of central, regional and municipal institutions, and on social networking websites. However, the most important thing for members of local communities is to record information about the flood they are experiencing 'here and now' and to use it in the process of creating local flood culture, sustaining the public memory of the event and finally internalising knowledge about how the local environment works. Any contents communicated to the public should relate to the local context as otherwise they are not understood or accepted by residents, have little positive effect on their own preventive activity, and do not lead to their acceptance of the planning and flood protection activities conducted by municipalities. This approach should be accompanied by the assessment of social vulnerability at a local scale. Even though studies such as those presented in this book show the general tendencies in flood preparedness as a result of certain social inequalities, local level studies could uncover a more complex structure of social vulnerability that depends on the context and situation. Various social aspects of flood risk management discussed in this book, chosen out of many interrelated topics within the field of disaster risk reduction studies, could be used as a support in strengthening resilience to floods and in implementing new paradigms of flood risk management, including a bottom-up approach, two-way risk communication and shared responsibility in flood risk mitigation.

## Reference

Tuan Y-F (2002) Space and place. The perspective of experience. University of Minnesota Press, Minneapolis